On-Site Occupational Health and Rehabilitation

On-Site Occupational Health and Rehabilitation

A Model for the Manufacturing and Service Industries

Jane Pomper DeHart

Department of Occupational Health
Henry Ford Health System
Detroit, Michigan

MARCEL DEKKER, INC. NEW YORK • BASEL

Library of Congress Cataloging-in-Publication Data

DeHart, Jane Pomper
On-site occupational health and rehabilitation : a model for the manufacturing and service industries / Jane Pomper DeHart.
p. cm.
Includes index.
ISBN 0-8247-8986-5 (alk. paper)
1. Employee health promotion. 2. Occupational health services. 3. Medicine, Industrial. I. Title.
RC969.H43 H485 2000
658.3′82—dc21
00-060194

This book is printed on acid-free paper.

Headquarters
Marcel Dekker, Inc.
270 Madison Avenue, New York, NY 10016
tel: 212-696-9000; fax: 212-685-4540

Eastern Hemisphere Distribution
Marcel Dekker AG
Hutgasse 4, Postfach 812, CH-4001 Basel, Switzerland
tel: 44-61-261-8482; fax: 44-61-261-8896

World Wide Web
http://www.dekker.com

The publisher offers discounts on this book when ordered in bulk quantities. For more information, write to Special Sales/Professional Marketing at the headquarters address above.

Current printing (last digit):
10 9 8 7 6 5 4 3 2 1

PRINTED IN THE UNITED STATES OF AMERICA

To my mother, who developed and survived breast cancer during the writing of this book. For her love and friendship.

And to my four nieces, Laura Dierwa, Kristin Dierwa, Larissa Pomper, and Melanie Pomper. May you always believe that dreams can come true and that you can leave a mark on society by using your gifts and talents to help others.

Preface

The magnitude of the occupational health burden can be made clear by comparing it to the costs of cancer and heart disease. Job-related injuries and illness result in costs to workers and employers that exceed those of cancer or heart disease. More than 13.2 million Americans are hurt each year in the workplace. Job-related injuries and illnesses have a direct cost of up to $178 million a day in the United States, according to reports from the National Institute of Occupational Safety and Health and the National Academy of Social Insurance. ''Reports further indicate that 20 percent of workers suffer at least one week of back pain annually. It remains the single most costly type of occupational health disorder in terms of lost wages, insurance claims and medical bills'' (Mahone and Burkhart, 1992). Low back pain alone costs businesses an estimated $40 billion a year. Back pain is second only to headaches in terms of the most frequently reported cause of pain in the United States (Mayo Clinic, 1994).

Many of the proposed solutions for containing these skyrocketing costs are beyond the control of an individual employer to shape, influence, or implement. Some examples of forces that employers cannot control independently are as follows: (1) Health maintenance organizations (HMOs) are combining group medical insurance with workers' compensation. (2) Various state legislatures are either considering or are actually passing legislation that fosters greater use of managed care by employers in order to restrain medical costs. (3) Insurers are competing to develop and implement the most successful approach to solve this problem.

The placement of on-site occupational health clinics and rehabilitation and fitness/wellness facilities in plants has been being written into local contract agreements. I have set up such services in the workplace of both private and publicly owned businesses and companies, and this book draws extensively on personal experiences, lending credibility to the recommendations given here. The companies at which these on-site services have been set up include Ford Motor Company, General Motors Corporation; Bank One Corporation, The Detroit Newspapers; J&L Specialty Steel Corporation, and the Henry Ford Health System. Off-site occupational medicine clinics have also been developed for the Pepsi-Cola Detroit Plant.

Individual employers are looking for solutions that they can put in place at their worksites now. These solutions must control costs, but they must also meet the real medical and rehabilitation needs of injured workers and their families. In this book I demonstrate that we can save significant amounts of money in the service and manufacturing industries by delivering full-service medical and rehabilitation programs on-site at the work location. This approach is straightforward and easy to understand, and it can be implemented in almost any setting, regardless of the industry or size of the employer.

The most crucial factor determining whether the on-site program will succeed is the culture operative at each client (i.e., company customer) worksite. In order to identify the characteristics of a worksite in which an on-site full-service medical and rehabilitation program will flourish, it is first necessary to identify the characteristics of the worksite where it does not: where workers' compensation costs are soaring and where the medical and psychological needs of the injured worker are not being well met. This worksite is paternalistic and has low trust, a hierarchical structure, task specialization, an adversarial context, structured communication, filtered information, and limited authority to act. On the other hand, the worksite in which the on-site approach flourishes is patient-centered and has high trust, a flattened structure, an informed communication process, cooperative relationships, and authority to act within appropriate parameters, and is patient centered (that cannot be said often enough). The implementation of an on-site program actually supports and encourages these positive characteristics.

The injured worker, the supervisor, and the union working together to facilitate early return to gainful employment will find this text useful. The model presented here serves as an orientation to corporate America's awareness of workers' compensation and the problems associated with lost work time, and it provides applied solutions.

Jane Pomper DeHart, MA, OTR

Acknowledgments

The on-site model of occupational health and industrial rehabilitation in the manufacturing and service industries would not have been successful without the significant contributions and collaborations of our on-site medical team. This includes personal communications, structuring the design, organizing and conducting statistical analysis, collecting data, and participating in "grand rounds"/ case management. I wish to acknowledge and thank the generous contributions of time and expertise provided by the following members of the on-site team during the past twelve years:

Julie LaTorella,† Alan Case, Thomas C. Royer, MD, Lucius C. Tripp, MD, MPH, Kim Kezlarian, MD, MSA, Carolyn Shettler, MD, Eleanor Stewart, MSA, OTR, Anthony Lewis, George Metropoulos, MD, MPH, Wally Gasiewicz, MD, MPH, Barbara Yakes, DO, MOH, James Fitzsimons, Rica Nazareno, RPT, Charles Zabinski, RPT, Thomas Lennon, RPT, David Moore, RPT, Alan Wryzkowski, RPT, Jane Darrow,† RN, Gary Schwabe, Nick Sekles,† Jim Dailey, Tim Fracassi, ATC, Scott Lessing, ATC, George Tsiminakis, Mary Dalton, RN, Walter J. Talamonti, MD, MPH, Debbie Stann, and Gregory Preston, MD.

I am also grateful to the workers, employers, and companies for whom the on-site services have been developed. The aforementioned assembly of healthcare colleagues has helped to build a model with great thoughtfulness, wisdom, and dedication to keeping those with workplace injury or illness healthy and productive.

A special thank you to Walter J. Talamonti, MD, MPH, Director, Clinical Operations, Ford Motor Company, for his technical editorship of this book.

†Deceased.

Contents

On-Site Occupational Health and Rehabilitation

1

The Mission and Vision of the On-Site Team

INTRODUCTION

The novel idea of the on-site, at-work model has evolved through the use of methods that enable injured workers to get back to work faster while saving money for the employer and cutting workers' compensation costs. Employers must pay for medical services and provide compensatory payment for lost wages as long as the employee is unable to work. This usually represents 80% of the worker's current salary or wage. By finding work that the employee is able to perform, the company reduces its liability and the worker gets better faster. Thus, the mission of the on-site occupational health and rehabilitation team is dedicated to continuously improving patient/employee care. The commitment is to design and implement occupational health programs for companies that have integrated group health, workers' compensation, and disability management for both occupational and nonoccupational injuries and illnesses affecting the employee's ability to perform in the workplace. The team's mission is to build a support system by remaining at the forefront of medical technology and rehabilitative health care delivery, preventing disease, promoting wellness, and thereby being responsible corporate citizens. This is accomplished by recognizing that criteria have been set for quality, price, and service.

To fulfill the mission, the vision is to work jointly with the company and its employees to minimize lost work time and enhance the health and safety of each worker, striving to keep employees working and prepare them for job placements that are appropriate to their impairments. This program gives employees both physical and mental support, without which, in the past, employees with disabilities faced barriers to continued employment. These goals are achieved through effective case management, return-to-work/transitional programs, occupational health physician networks, duration guidelines, and treatment protocols geared toward meeting quality outcome standards and compliance with state and federal regulations.

Implementation of other prevention technologies—such as back injury prevention protocols and rehabilitation programs to protect all workers from diseases and injuries—can reduce costs by cutting lost work hours and keeping productivity high. According to Daniel Mont in a study he conducted with John F. Burton, Jr., and Virginia Reno at the National Academy of Social Insurance, the cost to employers of workers' compensation in 1996 was $55.2 billion. This included premium payments and payments under deductible policies by employers purchasing insurance and benefit payments and administrative costs paid by self-insurers. It did not include any loss in productivity because of an injured workforce. These costs were $48.3 per covered employee and $1.67 per 100 of covered payroll. Improved quality of care provided to achieve benchmark status can be demonstrated. By interfacing information systems for work-related injury/illness outcome and tracking, employers can integrate with both health care providers and the corporations' guidelines, policy, and analysis. In order to help any management team cut workers' compensation costs, the on-site team works closely with the client, tailoring the program's design to each specific corporate structure, the corporate resources, and corporate goals.

For example, in the hospital business, prior to 1996, if a nurse who lifted a patient incorrectly sustained a back injury, the work-related injury or illness would be treated as if the nurse had a routine health problem. The employer and insurance company would have been required to use local existing clinical facilities and services for the nurse's treatment and workers' compensation care. Nurses with back injuries randomly entered the delivery system through the emergency department (ED), clinic, or hospital and received excellent traditional medical or surgical care. However, claims management, which is a process unique to workers' compensation care, was not well organized, resulting in poor processing and extreme dissatisfaction on the part of the primary client—the employer. Essential elements of workers' compensation care were not addressed, including work restrictions or status, follow-up care, duration of restricted activity, and effective communication by the physician to the nurse's supervisor in the department of nursing. A job placement was not explored or tracked for compliance on behalf of the injured nurse.

The lack of dedicated occupational health clinical and administrative expertise combined with a chauvinistic attitude among providers toward workers' compensation care can result in a poor commitment to care. Although a significant number of workers' compensation patients require more objective evaluation and attention than other patients with similar problems, there has been an exaggerated fear that these patients are all "symptom magnifiers" who are not motivated to return to work, and that it takes too much time to deal with their cases in terms of paperwork and the legal system. These attitudes and the lack of market pressure at the time result in a poor understanding of the financial opportunity workers' compensation care represents.

The best example of problems in dealing with occupational health issues is the approach taken in the past to employee health matters within the hospital setting, where there has been a lack of basic occupational health interventions such as injury-prevention programs, ergonomics, and aggressive placement programs. The industry standard for placement in a restricted work environment is 30 days. Hospital policy puts no limitation on the duration of temporary placement. In large hospitals (more than 900 beds), the monitoring of long-term cases has been poor, essentially resulting in the assignment of permanent disability to more than 200 Michigan hospital employees, some with minor disabilities, for which indemnity and health care costs have accrued for over a decade. The financial impact of ignoring these issues continues to exceed $5 million per year for large hospitals.

Delivery Models

The present author has had extensive experience in implementing workers' compensation care and industrial rehabilitation programs since 1989. Each arrangement differs in the range of services it provides, demonstrating flexibility in response to customer needs.

Model I: On-Site Industrial Rehabilitation

The Henry Ford Health System/General Motors (GM) On-Site Industrial Rehabilitation Program serves as a nationally recognized model for providing comprehensive case management. The program, in its tenth year of operation, is designed around the needs of the client. It has served 19,000 GM employees at the North American Operations (NAO) Technical Center in Warren, Michigan, and approximately 3300 at the Orion, Michigan, Assembly Plant. The author worked with both Henry Ford Health System and General Motors Corporation legal departments to design a direct company contract for on-site occupational therapy hand-treatment programs (i.e., carpal tunnel syndrome, de Quervain's disease, tenosynovitis, mallet deformities, finger amputation, etc.) and physical therapy (back problems, epicondylitis, knee injuries, thoracic outlet syndrome, etc.).

In 1991 GM contracted with the Henry Ford Health System to open a second facility in Michigan at the GM Saginaw Steer & Gear Plant (now an automobile parts supplier plant). The NAO Technical Center program was duplicated at the Saginaw plant and then expanded to Detroit in 1994. In 1997 a fitness/ wellness and rehabilitation facility and a transitional work reconditioning center was opened at the GM Orion facility, where cars are assembled from start to finish. In 1997 the GM Lansing (Michigan) Assembly Plant agreed to place a certified athletic trainer on the production floor to anticipate injuries before they occur, assessing poor body mechanics or improper lifting techniques and their effects on the back or shoulder. The present author began her work in American plants by applying safe-workplace principles of stretching, strengthening, and body positioning. Her toughest obstacles lay in recruiting and retaining industrial rehabilitation staff who were capable of working in plants remote from hospital or clinic environments. In every setting, appropriate isokinetic equipment (i.e., Lumbar MedX treatment) and modalities (heat, cold, electrical stimulation, ultrasound, iontophoresis, etc.) were purchased and brought on site for the clinicians' use. Cervical lumbar traction, customized hand splints, myofascial release, and trigger point release are examples of other specialized musculoskeletal techniques and treatment provided on site in the plants.

Model II: On-Site Rehabilitation and Medical Services

J&L Specialty Products Corporation (J&L Steel) contracted for an on-site board-certified occupational medicine physician in 1992 to evaluate, diagnose, and treat patients and return employees to work. Clinical space had been provided by the company for an on-site physician as well as physical and occupational therapy services. A local hospital was selected as the site for urgent care, testing, and 24-hour emergency department services. Occupational health staff members work as a team with workers' compensation representatives at the plant. The agreement includes a transitional work program. The Detroit Newspaper Agency is another client. The on-site services include one board-certified occupational medicine physician, one physician assistant, and 8–14 nursing support staff at two on-site locations. The Detroit Newspapers require Monday-through-Sunday three-shift operations for medical staffing. To date the largest plant serviced is one supplying automobile parts; it has three shifts of 36 medical staff who are on site Monday through Sunday.

Model III: All Services Provided at Off-Site Facilities

These services are given to a small manufacturing employer with two plants in close proximity to local medical centers in Troy and St. Clair Shores, Michigan. Employees receive preplacement evaluations, workers' compensation evaluation and treatment, and emergency care at local facilities, with no on-site care provided. The author was able to contract for several small- to medium-size businesses to utilize off-site facilities due to the lack of space on site. A central office case manager manages the referral. The Detroit Pepsi-Cola plant also uses the off-site model. Pepsi is located approximately 3 miles from the Detroit Occupational Health and Industrial Rehabilitation facility.

Model IV: Mixed Services

Bank One Corporation supports two on-site clinics for both medical and rehabilitation services at its southeastern Michigan processing centers. Employees from 150 branches receive workers' compensation care at a local medical center throughout the delivery area. It should be noted that on-site clinics can be set up for small (200 employees) or large (19,000 employees) operations in either the service or the manufacturing industries.

Current Status

The author's occupational health and industrial rehabilitation service now manages over 46 delivery relationships with employers, including work with international companies, direct company contracts, letters of agreement, and preferred provider arrangements. Building a successful on-site occupational health and industrial rehabilitation practice relies on developing relationships with clients, understanding their needs, and providing services to meet those needs. Knowing the customer has a significant impact not only on how your evaluation and treatment programs are structured but also on the kinds of service programs to be developed in the future. Clients include unions, benefits representatives, management, physicians, and therapists, who monitor quality assurance, outcome, patient protocols, measurable rehabilitation goals, and statistical data collection.

ACCESS EQUALS LOCATION

Employees no longer have to wait 4–8 weeks for access to local clinic or hospital facilities. The on-site program eliminates off-job time that employees formerly took to travel to and from therapy or medical treatment. The appointment is scheduled for the same day, the next day, or the following day (within 72 hours) so that employees are rehabilitated quickly. The on-site physical and occupational therapy program meets the standards of the Joint Commission on Accreditation of Healthcare Organizations (JCAHO) and the elements of ISO-9000. Becoming ISO-9000–registered means that the provider of occupational health care will meet the automotive quality standards that the ''Big Three'' are requiring their health care suppliers to meet. Employers are recognizing the importance of controlling health care costs.

The unique location of industrial rehabilitation on site gives therapists a chance to monitor an employee's condition directly by evaluating and observing the work station. The therapist can structure therapy around the patient's actual work environment and create an appropriate conditioning program to help minimize lost work time as contrasted to a work reconditioning program in a clinical center (off site). Also, the location facilitates communication between on-site company physicians. The physician, the nurse, the therapist, and the ergonomist—working together with the patient, the supervisor, and the union—all become part of the rehabilitation team and part of the solution to the problem. Therapists have pulled workers right off the assembly line to correct improper posture. This has cut the number of injuries by 53% in the last 12 years. Cost savings for one of our companies have totaled over $7 million since the program was initiated.

The objective is continuous quality improvement by striving to keep employees fit and well through prevention and proper reconditioning. Reconditioning involves keeping the injured/ill worker at work and gradually increasing the amount of repetitive work and the duration of the tool or job task until the worker can sustain and endure his or her full job duties, resuming work with significantly reduced pain and symptoms. The amount of reconditioning is tailored to the individual worker's needs. Guidelines have shown a period of 30 days or less to be effective in the majority of cases. Every day the focus is on prevention and early intervention. Healthy living is demonstrated by monitoring usage in the fitness center and by providing educational materials on back problems and high blood pressure. Health-promoting wellness fairs are held at least annually. Employees are encouraged to undergo glucose testing, body fat composition analysis, vision evaluation, blood pressure checks, and grip strength tests, among others.

On-the-job injuries hurt in more ways than one in that they cause pain to employees and serious losses to employers. With more than 5 million workers injured on the job every year, the author has three goals: to reduce workers' compensation costs, return employees to work sooner, and reduce the number of injuries, including cumulative trauma injuries. At J&L Specialty Steel, Inc., in Detroit, the author introduced a comprehensive program that includes claims management, on-site occupational health care professionals, safety and injury-prevention classes, workplace and ergonomic analysis, and the development of temporary modified work programs. As a result, J&L has had a savings of over 60% in workers' compensation costs from lost work time. ''Since we began the program we've realized a tremendous cost savings in lost man-hours alone,'' said George Tsiminakis, manager of Industrial Relations. ''One area of the program that really works for us is having health care professionals on site for assessment and treatment of injuries. It leads to quick recovery, reduced claims costs, and a smooth transition back into the workplace. On-site services are one of the keys to the program's success. Bringing health care to the place of employment reduces barriers to recovery.'' Another key is the claims-management process, which involves a joint company and health team that reviews each claim and monitors quality assurance outcome, patient protocols, and measurable rehabilitation goals. "Ultimately it gets our workers back to their jobs sooner," said Robert Salter, safety supervisor at J&L Specialty Steel.

Because preventing injuries is just as important as treating them, the on-site occupational health program includes on-site ergonomic job assessment and safety awareness assistance. Approximately four out of five Americans suffer from low back pain, which has become the most common cause of disability for persons under the age of 45. Table 1 lists common rehabilitation workers' compensation diagnoses in an on-site automobile manufacturing company during a 6-month period.

Companies, especially in the manufacturing industry, have been interested in the cost and incident rate of work-related injuries and illnesses because of the need to keep productivity high. Replacement workers are also an added cost. The employer is paying for both the workers' compensation liability and the replacement worker. Workers' compensation laws were enacted between 1911 and 1914. Through the years, there have been employer-sponsored clinics, but

TABLE 1 On-Site Diagnostic Visits for Rehabilitation

Diagnosis	Number of Visits
1. C7-6 radiculopathy	5
2. Cervical and lumbar strain	15
3. Cervical radiculopathy/arthritis	8
4. Cervical radiculopathy	8
5. Cervical thoracic strain	8
6. Degenerative joint disease with right-arm radiculopathy	7
7. Adhesive capsulitis, shoulder	21
8. Low back pain	6
9. Back strain	5
10. Cerebrovascular accident	12
11. Lower extremity radiculopathy/cerebrovascular accident	5
12. Lumbar spondylosis	12
13. L2-L5 sciatica	12
14. Carpal tunnel syndrome	4
15. Neck/shoulder strain	10
16. Shoulder strain	6
17. Elbow pain	7
18. Degenerative joint disease, low spine	7
19. Lateral epicondylitis	8
20. Low spine spasms	8

TABLE 1 Continued

Diagnosis	Number of Visits
21. Status post–ruptured cervical disc	8
22. Severe degenerative joint disease/lumbosacral pain	8
23. Chronic lumbosacral strain	9
24. Neck pain	10
25. Status post–L5 discectomy/low back pain	8
26. Cervical strain	6
27. Left shoulder tendonitis	19
28. Chronic leg pain	2
29. Chronic neck pain	9
30. Anterior cruciate ligament/meniscus tear	7
31. Back pain, paraspinal spasms	3
32. Low back pain	11
33. Status post–repair to anterior ankle tendon	20
34. Status post–malleolus ankle fracture	10
35. Status post–laminectomy with foot drop	3
36. Rheumatoid arthritis (hand)	2
37. Right knee contusion	17
38. Recurrent ankle sprain	7
39. Recurrent dislocated shoulder	16
40. Severe paravertebral trapezius spasm	12
41. Degenerative joint disease/reversed lumbar lordosis	9
42. Lumbosacral strain and sciatica	10
43. Peroneal tendon sprain with foot pain	2
44. Whiplash/lumbar radiculopathy	8
45. Multiple sclerosis/leg weakness	16
46. Subluxation right shoulder	4
47. Severe contusion bilateral thighs	12
48. Back pain	11
49. Lumbosacral muscle strain	5
50. Paravertebral trapezius spasms	5
51. Anoxic encephalopathy	22
52. Acute cervical myelopathy	7
53. Bilateral small infarction/cardiac disease	8
54. Left sciatica	4
55. Lumbar muscle strain	10
56. Status post–knee arthroscopy	8
57. Lumbosacral sprain	8
58. Lumbar radiculopathy	9
59. Shoulder injury/possible rotator cuff tear	14
60. Shoulder tendonitis	12
61. Lumbosacral strain	8
62. Lumbosacral strain	7

Table 1 Continued

Diagnosis		Number of Visits
63.	Osteoarthritis, both knees	2
64.	Spasm and pain status post–motor vehicle accident	13
65.	Cervical strain/shoulder tendonitis	8
66.	Low back pain	8
67.	Impingement right shoulder/status post–cuff repair, left	9
68.	Closed head injury	3
69.	Torn gluteus maximus	6
70.	Lumbar intravertebral disc/myelopathy	6
71.	Cervical radiculopathy	8
72.	Lumbosacral radiculopathy	2
73.	Right shoulder pain	11
74.	Right lumbar disc herniation	4
75.	Repaired torn ulnar collateral ligament, left thumb	23
76.	Sciatica, left leg	9
77.	Reflex sympathetic dystrophy	8
78.	Cervical sprain	2
79.	Low back pain with radiculopathy	8
80.	Degenerative disc disease with sciatica	8
81.	Chronic back pain	5
82.	Right L5 radiculopathy	9
83.	Persistent neck pain	3
84.	Cervical radiculopathy	11
85.	Low back pain and right shoulder pain	8
86.	Lumbar radiculopathy	12
87.	Muscle contraction headaches	4
88.	Left shoulder tendonitis	8
89.	Neck pain radiating to left shoulder	18
90.	Low back pain	11
91.	Myofascial syndrome of left trochanteric area	6
92.	Lateral meniscus tear	8
93.	Cervical radiculopathy	15
94.	Recurrent low back pain	8
95.	Neck pain, muscular	10
96.	Resolving sciatica	6
97.	Low back pain/radiculopathy	19
98.	Cervical radiculopathy	8
99.	Torticollis secondary to motor vehicle accident	11
100.	Inferior patellar tendonitis	10

Source: The most commonly used classification of injuries for workers' compensation claims, the *International Classification of Diseases*, 9th rev. These codes were applied on site at an automobile manufacturing company during a 6-month time frame.

an on-site model including a full-service program of physician services, nursing, physical and occupational therapy, transitional work, ergonomics, and fitness/prevention was not as readily realized.

The on-site team builds relationships in tune with the employer's culture. The philosophy of the plant operation, the production system, and the competitive manufacturing principles are factors that the provider of health care is exposed to and works to improve by keeping the employee/patient well. Direct company contracting with purchasers is not easy to achieve. Contracts are usually awarded for a period of 3 years, and the contract for on-site services has volatility. For example, the union may restrict workplace practices, so that there may be no freedom to assign work outside of the normal routine of a job description without the union's involvement. Seniority may contribute to a verdict of ''no job available'' when attempts are made to place an injured worker with or without a restriction back on the assembly line. The company or plant may be steeped in politics. Elections are held for a bargaining committee, where the efforts and actions driven by mutual gains for the membership may conflict with the need to run the company efficiently and competitively. The health care providers must work within a dual salary/management and union structure. Factory rules may be outdated. The ultimate goal of both the union and management is to maintain high-quality, outcome-driven decision making. The injury prevention and loss control program is intended to build strength and the ability to tolerate repetitive work. Tool use, safety, and ergonomics—for example in a ''paint booth'' area or on a ''seal gun'' assembly line—will require that the worker have good flexibility, fine motor coordination, proper body mechanics, and physical conditioning. The union, management, worker, and health care provider must build a relationship fitting the author's on-site model. By working together, health and safety in the workplace are achieved.

The Occupational Safety and Health Administration (OSHA) of the U.S. Department of Labor keeps statistics on the incident rates of the number of injuries and illnesses per 100 full-time workers in the agricultural, mining, construction, manufacturing, transportation, wholesale, retail, financial, and services industries. Table 2 presents these statistics for the manufacturing industry.

The Agreement for Occupational Medicine Services by and between the provider and company includes a central toll-free telephone number for the coordination of services. Requests for Occupational Health and Rehabilitation Services may be written by an employer in the form of a Request for Proposal (RFP) or a Request for Quotation (RFQ). Usually the bidder or provider will have 4 days to 2 weeks to respond to the RFQ. Appendix 1 is a sample of an automobile company's request for plant on-site occupational medicine physician services. The request may be for on-site or off-site services, for rehabilitation (physical therapy, occupational therapy), for fitness, wellness, prevention, ergonomics, nursing, and/or board-certified occupational medicine physician services. Typical

Table 2 Manufacturing Industry Incident Rates[a] of Nonfatal Occupational Injuries and Illnesses, 1997

Industry[b]	SIC[c]	Injuries and illnesses, total cases	Injuries and illnesses, LWDI[e]	Injuries, total cases	Injuries, LWDI[e]
Manufacturing	20	10.3	4.8	8.9	4.2
Durable goods	200	11.3	5.1	9.8	4.5
Meat-packing plants	2011	32.1	18.7	18.7	10.5
Cereal breakfast foods	2043	8.1	3.9	7	3.6
Bakery products	2050	11.2	6.3	10.3	5.8
Tobacco products	2100	5.9	2.7	5.6	2.7
Lumber and wood products	2400	13.5	6.5	12.8	6.2
Paper mills	2620	6.2	2.9	5.6	2.7
Tires and inner tubes	3010	12.2	7.3	11.2	7
Plastic bottles	3085	10.3	5.7	9.9	5.5
Steel foundries	3325	19.1	10.4	18.3	10
Sheet metal work	3444	16.2	7	15.5	6.5
Iron and steel forgings	3462	17.3	7.7	15.7	0
Construction and related machinery	3530	13.9	5.7	13.1	5.4
Electronic computers	3571	2.3	1.2	1.6	0.9
Motors and generators	3621	9	4.2	7.1	3.2
Aircraft parts and equipment	3728	9.9	4.5	8.3	3.6
Shipbuilding and repairing	3731	21.4	10.7	18.3	9.6
Manufacturing industries	3999	8.9	4.8	8.4	0

[a] The incidence rates represent the number of injuries and illnesses per 100 full-time workers and were calculated as

$$(N/EH) \times 200{,}000$$

where

N = number of injuries and illnesses
EH = total hours worked by all employees during the calendar year
200,000 = base for 100 equivalent full-time workers (working 40 hours per week, 50 weeks per year).

[b] Totals include data for industries not shown separately.

[c] *Standard Industrial Classification Manual*, 1987 ed.

[d] Employment is expressed as an annual average and is derived primarily from the Bureau of Labor Statistics State Covered Employment and Wages program. Employment for private households (SIC88) is excluded.

[e] Total lost-workday cases involve days away from work, days of restricted work activity, or both.

Source: The Occupational Safety and Health Administration, U.S. Department of Labor, statistics on incident rates of the number of injuries and illnesses per 100 full-time workers in the manufacturing industry.

questions asked by manufacturing companies are delineated in Appendix 1. The sample RFP was written to address the most common occupational medicine practices provided in a plant setting. However, a city, municipality, or service industry may ask very different questions. The city worker may work off site in the public works department, or the worker may be a police officer or firefighter, whose job duties are not performed in one location. For example:

> *How does your organization plan to help—for example, a city, municipality, or corporation—reduce workers' compensation costs? Please address in detail.*

The answer may be to state that services for the city or municipality workers' compensation program will be coordinated through a central office/intake number, which would coordinate cases from the initial telephone call through the completion of care and ultimately the case closure. The occupational health program is integrated and has the ability to provide most services within its organizational structure, therefore decreasing referral and medical record transfer time. Cases can be facilitated through on-line data transfer, accessible to physicians and therapists. The central office will also have on-line access to appointment scheduling/locations, which will facilitate the services to city employees. The team implements workers' compensation programs so that employees encounter no waiting time when they seek treatment in on-site clinics (i.e., police headquarters), hospital EDs, or off-site medical centers. All of these attributes help to reduce the costs associated with work injuries by decreasing the time to the initial visit as well as the time between appointments. Restrictions such as "no use of vibratory tool" or "one-hand job" are monitored every 4–7 days. An anticipated date of recovery is documented. An injured worker who is unable to return to the same job may be placed in a temporary position for up to 30 days. A permanent restriction may be written for extensive injuries (i.e., right dominant hand amputation of four digits). Job placement is based on seniority and job compatibility or capability. The union plays an important role in job placement. If no job is available, retraining is suggested (i.e., "dispatcher job" or educational training for computer sit-down work, etc.). In 2–3% of cases, a settlement is offered to the injured/ill employee in order to close the case. The settlement, under legal sanction, may bestow a sum of money, which is usually an amount that the injured/ill worker believes to represent fair to secure income. An agreement between the worker and the company is arrived at through mediation.

On-site providers currently service workers' compensation third-party administrators as well as other insurance programs such as health maintenance organizations, preferred provider organizations, point of service, and traditional plans. In terms of workers' compensation, the author promotes preventive care, review of return-to-work orders, and training at the worksite to decrease the potential of work injuries. Again, these preventive measures decrease the incidence of work injury, thus decreasing the cost of medical services, therapy, and lost work time. See Chapter 3 for a sample of the agenda for worksite injury prevention education. Managing workers' compensation care can take the form of pricing strategies that include three separate choices, as defined in Chapter 3, such as a *case rate*, all-inclusive *monthly management fee*, or *discounted rate* for direct company contracts. Low cost, high quality, and client satisfaction are measured and reported directly to the company's contracted client in all three pricing methods.

What training or experience do the physicians have in occupational medicine?

The physicians have several years of experience in occupational medicine program development within the specialty area of workers' compensation injury/illness treatment, workplace design, and outcome measurement. Members of the delivery team of physicians may have a master's degree in public health and are board-certified in occupational medicine by the American Board of Preventive Medicine. Certification in the treatment of alcohol and drug abuse has also been obtained by the physicians. The key to success is the treating physician, who determines when the employee may return to work. An identified medical delivery team of physicians provides care. For example, the occupational medicine physician informs the hand surgeon of the physical requirements of the employee's job. All information is presented at a multidisciplinary team-care meeting (grand rounds and/or case management), with the delivery team of physicians present to review each case and supporting documentation and to make a disability duration-of-treatment recommendation.

Specialty physicians are available for independent or second opinion medical exams (i.e., back and spine clinic and neurosurgery services). These services include coordinated care, active management, early intervention, and physician communication by the expert occupational medicine provider team. A closed-loop system of approved physicians is utilized. The occupational medicine physicians serve as the gatekeepers.

What portion of the physician's practice is devoted to workers' compensation?

All of the occupational medicine physician's practice is devoted to workers' compensation. For more severe problems—such as cut ligaments, cut tendons, or back disease, the occupational medicine physician expert will utilize the credentials of a network of board-certified specialty providers. These providers have experience in writing appropriate, time-limited restrictions. Follow-up appointments are made weekly or every 2 weeks to monitor the progress of injured/ill workers. Timely documentation in the medical record for the employer is provided, and the employee's supervisor receives a new copy of the activity prescription (Fig. 1) guideline following each visit.

NAME:____________________

OCCUPATIONAL HEALTH **Medical Record Number:**_______________

SPECIALTY PHYSICIAN

ACTIVITY PRESCRIPTION

DIAGNOSIS:___________________________ Date of Service___________________

The following medical activity guidelines are recommended:

- ☐ Full activity/no guidelines __________
- ☐ No activity/rest of shift only_________
- ☐ No activity/unable to work__________
- ☐ Activity modifications______________
  - ☐ No lifting over ____ pounds
  - ☐ No excessive bending
  - ☐ No lifting below knee/above shoulder
  - ☐ Seated work only
  - ☐ No prolonged standing/walking
  - ☐ Ground-level work only
  - ☐ No climbing/avoid stairs
  - ☐ Avoid kneeling/squatting
- ☐ Limited push, pull, reach right/left upper extremity
- ☐ No use right/left upper extremity
- ☐ Avoid use of vibratory or power hand tools
- ☐ No work at or above shoulder level
- ☐ Keep involved areas clean & dry
- ☐ Avoid skin irritants
- ☐ Remove from____________________
- ☐ Clean atmosphere; avoid dust, smoke, fumes, mist
- ☐ One eye work only; must wear eye patch
- ☐ Must wear protective eye gear
- ☐ No hazardous machinery or industrial vehicle driving
- ☐ Limited physical exertion

Figure 1 The activity prescription form, used to document the capabilities of the injured/ill worker.

- ☐ Limited grip & squeeze right/left hand & wrist
- ☐ No repetitive use right/left hand & wrist
- ☐ No palm buttons/palm pushing
- ☐ Job rotation recommended
- ☐ Limited shift to ____ hours
- ☐ May use air/vibratory tools with torque < 20 nM
- ☐ Other________________________________

Length of recommended activity

Return appointment:_______________ **Physician signature**____________________

Return to work date:_______________

Comments:

PATIENT: **RETURN THIS FORM TO YOUR EMPLOYER IMMEDIATELY.**

PHYSICIAN: **FAX THIS FORM TO THE OCCUPATIONAL HEALTH DEPT**

(313) 874-6037 FAX

MEDICAL RECORD, WHITE

EMPLOYER, PINK

THIRD-PARTY ADMINISTRATOR, YELLOW

EMPLOYEE, GOLDENROD

Figure 1 Continued

Are occupational nurses included on the staff? If yes, what is their role?

Occupational nurses are included in the staff. Nurses complete the OSHA log and all necessary benefit program forms. Occupational health nurses assist the physician with treatment procedures, the design of program policies and procedures, and education and training (e.g., regarding bloodborne pathogens, hepatitis B vaccine, compliance monitoring). (See Appendix 2, which notes the occupational health nurse's job description.)

Describe the clinic's patient follow-up tracking system.

The patient follow-up tracking system consists of a centralized office that coordinates the processing of employee/patients and paperwork, which includes intake or patient registration, diagnosis, date of injury, lost work time, nature of the injury, evaluation, treatment referral, scheduling, reporting and billing, and case management through the use of a toll-free telephone number. There are dedicated product line managers (both physicians and administrative personnel) working through the centralized occupational health and industrial rehabilitation office to ensure consistency of care by all providers. The centralized office of occupational health and industrial rehabilitation has developed and handled the delivery of medical care and physician management to reduce variations in clinical practices. ISO-9000 registration and the JCAHO helps the team focus on how to continuously improve and to set up and manage its processes in order to meet specifications for its occupational health products and services. The establishment of processes and process control for patient/employee care is the aim. This includes credentialing, contracting/hiring, management reporting, protocols of care, sites of care, purchasing, case management, and referral to outside providers (e.g., an ED). The program ensures efficient delivery of services and appropriate closure of cases; its clinicians/providers have experience, expertise, and the ability to tailor programs to the needs of employers/insurers. An automated data system for results tracking and utilization is used to evaluate progress. Customer satisfaction surveys are utilized for feedback on the treatment rendered, timeliness of services, and outcome. (See sample of patient satisfaction surveys in Chapter 4.)

Critical success factors include a reduction in lost work time. This is specifically achieved through a single point of access for clients (e.g., city workers), for whom services are provided at convenient off-site clinics using centralized telephone intake and initial coordination of the arrangements for care. Utilizing established protocols, the patient will be directed to or given an appointment for evaluation. Referrals for central intake will be initiated by the company (e.g., the city) or the workers' compensation disability insurance carrier. Workers' compensation patients who access emergency departments or clinics directly will be identified and referred to the central intake area for coordination of care and follow-up activities after initial services have been provided. For customer companies that have on-site facilities, care will resume at the on-site location. In some states employees are treated by the on-site physician provider during the first 10 days following the injury/illness. After the first 10 days, the employee may work with any provider of his or her choice. If a good experience and rapport develops between the company physician and the injured/ill employee, the case is more frequently closed. The worker, the employer, and the medical team are all aware of the job requirements and the worker's capabilities for the job duties.

CASE EVALUATION

Case evaluation serves as the initial and ongoing source of illness/injury treatment planning. Services include initial follow-up from emergency department treatment or the first on-site scheduled assessment. The patient receives a physical assessment (i.e., musculoskeletal evaluation, test of respiratory fitness, and drug screen), the necessary diagnostic studies are ordered and reviewed, and a course of treatment is prescribed and monitored. The circumstances of the illness/injury are evaluated to determine if the case is work-related or the result of some other cause (e.g., a systemic malignancy). The case evaluation guides and promotes an active return-to-work plans, making determinations as to the extent of restrictions if any. Expert evaluation or an independent medical exam (IME) is included as required. This exam is usually a comprehensive 2- to 4-hour examination by a third-party board-certified specialty physician. The IME may be beneficial in litigation cases or when there is a conflict between what the worker believes he or she is able to perform and what the company physician reports.

CASE COORDINATION

Case coordination includes a number of activities, both administrative and clinical, that support the program goals, assist in regulatory compliance, and streamline processes for treating clinicians. Activities include creating a dedicated account, documentation and reporting, and focused customer support through enhanced communication. The process flow chart in Figure 2 depicts the on-site occupational health management process for workers' compensation cases.

CASE REVIEW—UTILIZATION REVIEW/ QUALITY MANAGEMENT

Consistent with the JCAHO and other performance standards, the managed workers' compensation program will implement utilization management and quality management programs. Using industry standard criteria, clinical protocols/guidelines, and clinical judgment, utilization of services will be evaluated for appropriateness, cost-effectiveness, and benefit eligibility. Individual cases will be re-

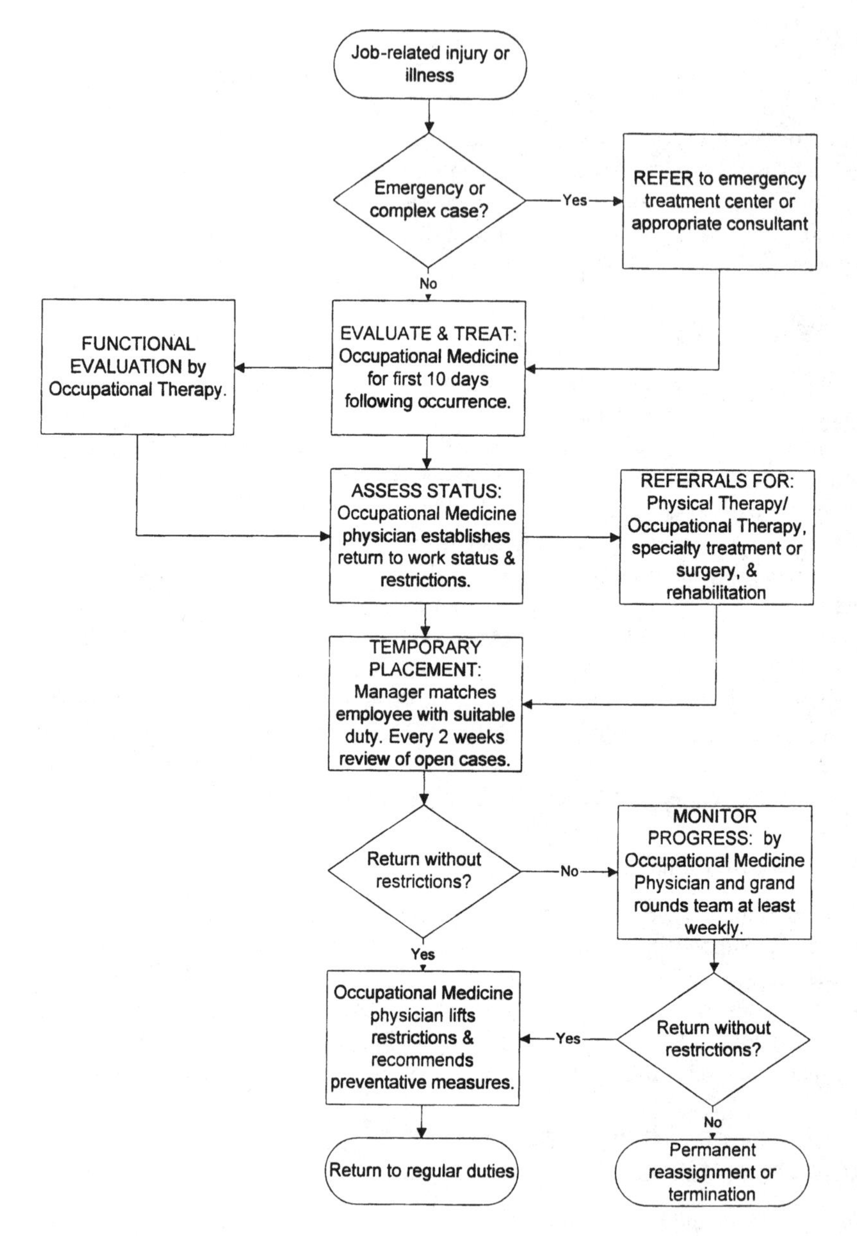

FIGURE 2 A management process developed by the author and widely used on site to deliver workers' compensation care.

viewed both concurrently and retrospectively to determine the best approaches to care management and to identify opportunities for improvement and improved case closure rates. ISO-9000 registration, used as a tool, has assisted the occupational health team in developing the quality management processes and documenting control in order to meet the goals of low cost, high quality, and customer satisfaction within the on-site clinic model and the patient/employee care plan.

What is the clinic's time standard for sending out a doctor's first report? Is this monitored?

The standard is 24–48 business hours for off-site primary occupational medicine reports to be dictated, transcribed, and sent to the employer, although current information systems can digitally dictate reports in final form 3–12 hours after a patient encounter. In the event that an employee is treated in the ED, a clinic/ED memo is created, which includes clinical notes and the return-to-work order. If this format is acceptable, as it has been for other employers as well as workers' compensation carriers, it can be electronically retrieved on site at the company by the on-site provider. However, in many of the author's on-site services, the automated medical records have been interfaced within the company and hospital to enable the on-site provider to pull up the computerized automated report from the hospital's laboratory or to read the finding of magnetic resonance imaging (MRI).

Briefly describe the physician assistant/nurse practitioner monitoring program.

Physicians review individual cases on a daily basis. The physician assistant may perform the initial evaluation of the patient to elicit a detailed and accurate history, perform an appropriate physical examination, and record the present pertinent data, including interpretive recommendations, in a manner meaningful to the physician. The latitude that the physician assistants/nurse practitioners are given in making diagnoses, ordering tests, and ordering prescriptions is always followed up by the attending occupational medicine physician. The physician assistant may perform or assist in the performance of routine laboratory and related studies as appropriate for a specific setting (blood studies, urinalysis). The physician assistant may accurately and appropriately transcribe or execute standing orders and specific orders at the direction of the supervising physician. (See the physician assistant job description, Appendix 3.)

How were the specialists selected and where are they located?

Specialists are identified by the treating occupational medicine physician, who orders the consult for the respective specialty discipline (e.g., plastic surgery for carpal tunnel release). Some cases require a telephone call to the central office by the specialist to obtain the diagnosis, course of treatment, medication, date the symptoms first appeared (or the accident happened), and the patient's progno-

Referral Form

DEPARTMENT OF OCCUPATIONAL HEALTH

Contact person:_____________ Phone number:__________________________

(Please fill out patient information)

Patient____________________Date of birth_________Medical record number______

SS#_______________Sex___Male__Female occupation__________

Mother's maiden name____________________Marital status __S __M __W __D

Employer__________________Work phone_________Home phone_________

Home address____

City______________________State__Zip__________

Date of injury_______________Diagnosis_____________________________

Purpose of referral:

- ☐ Workers' compensation evaluation and treatment
- ☐ Medical review officer
- ☐ Department of transportation (DOT) physical
- ☐ Non-DOT physical
- ☐ Drug screen collection only
- ☐ For cause drug screen
- ☐ 5-Panel drug screen test ☐ 6-Panel ☐ 10-Panel
- ☐ Breath alcohol test (BAT)
- ☐ Executive physical
- ☐ Follow-up appointment to discuss lifestyle changes
- ☐ Ergonomics/prevention
- ☐ Total & permanent disability Evaluation

- ☐ Preplacement exam with health history
- ☐ Latex sensitivity screening
- ☐ Lab: Latex-specific IGE
- ☐ Purified protein derivative (PPD) TB test
- ☐ Chest x-ray
  PA_____PA&LAT_________
  "B" reading_______
- ☐ Lab antibody titers
  Rubella_____
  Rubeola_____
  Mumps_____
  Varicella_____
  Hepatitis B_____

FIGURE 3 Department of occupational health referral form.

☐ Independent medical evaluation (IME)

☐ Surveillance exam: type__________

☐ Respirator exam

☐ Spirometry

☐ Vision screening

Titmus_____

Laser exam_____

Tonometry_____

☐ Lab tests

Prostate-specific antigen_____

Thyroid (TSH)_____T4_____

Biochem profile_____

Occult blood_____

Lead testing_____

Complete blood count (CBC) with differential_____

Urinalysis_____

Other____________________

☐ Vaccines

Measles, mumps, rubella_____

Hepatitis B (3 doses)_____

Tetanus_____

Hepatitis A_____

Other__________________

☐ Audiogram

☐ Electrocardiogram (ECG)

☐ Other

Insurance information:

____

____

____

____

Figure 3 Continued

sis for return to work. The occupational medicine physician will make a determination for returning the employee to a light or modified job where appropriate. All information from a specialist is presented to the occupational medicine physician team, which reviews each case and supporting documentation, including lab or test results, and measurable therapeutic goals to make a disability duration recommendation.

What procedures are in place to ensure that injured workers are referred to physicians within the preferred network?

All work-related injuries should be handled through a central office, which has a toll-free number for employer/employee access. The personnel at this office are trained in case management and appointment scheduling. It is the policy to utilize the office's own extensive delivery system, thus ensuring that the injured worker remains within the preferred network of physicians who can provide high-quality and expedient workers' compensation care. In the event that an employee is initially seen by a nonparticipating provider, the central office will coordinate the transfer of the case given the employer's approval. The employee is informed of the company choice for handling the workers' compensation claim. Procedures should be in place to assure that injured workers are referred to physicians within the preferred network. An ''intake sheet'' and authorization form initiated from the central occupational health department to the preferred specialty network provider should be included. Telephone communication should be ongoing. Figure 3 shows a typical department of occupational health referral form.

Briefly describe the treatment protocols for strain injuries (e.g., ankle). At what point, and based on what tests, is physical and occupational therapy begun?

Treatment protocols for strain injuries include clinical competency, history taking, evaluating the stress injury to the muscle/tendon (stretching of fibers and bleeding may occur), and evaluating whether ligament and joint structures are still essentially intact. The strain management protocol for muscle is shown in Table 3.

Physical and occupational therapy can be ordered and started within the first 24–48 hours. Typically, modalities are ordered: range-of-motion exercise and weight bearing as tolerated. Mobilization, progressive exercise, and a home exercise program are issued. Reassessment is provided and a physician progress note is given. Therapeutic modalities include the use of ice followed by heat. Pain and edema reduction may speed the healing. The length of patient/worker treatment is based on the degree of involvement and functional loss of the extremity. First-degree injuries may be accompanied by minor symptoms. Again, ice over the injured muscle may be utilized. In patients with second-degree strains, the injured muscle may be immobilized, the extremity elevated, and ice packs applied for the first 24–48 hours. Following this, the muscle should be ''at rest''

TABLE 3 Strain-Management Protocol for Muscle Injuries

1. First-degree strain (mild strain or slightly pulled muscle)
   a. There is trauma to a portion of the musculotendinous unit from excessive forcible use of stretch. Assess range of motion, strength, edema, tissue, pain, sensation, gait, and stability.
   b. Symptoms include local pain aggravated by movement or by tension of the muscle itself. The patient/worker may experience mild spasm, edema, ecchymosis, local tenderness with palpation, and a minor loss of functional use and strength.
2. Second-degree strain (moderate strain or pulled muscle)
   a. The patient/worker may experience trauma to a portion of the musculotendinous unit by extreme force or sudden intense activity (an excessive forcible stretch may be exhibited).
   b. Symptoms can include local pain that increases with movement or tension of the muscle, moderate spasm, edema, ecchymosis, and impaired functional muscle use.
   c. Tearing of fibers without complete disruption may occur.
3. Third-degree strain (severe strain or pulled muscle)
   a. Symptoms include severe pain and lack of full activity, severe spasm, swelling, ecchymosis, hematoma, tenderness, loss of muscle use, and usually a palpable defect.
   b. X-ray can demonstrate an avulsion fracture at the tendinous attachment as well as soft tissue inflammation/edema.
   c. Ruptured muscle or tendon; resultant separation of muscle from muscle, muscle from tendon, or tendon from bone may occur.
   d. Treatment may include immobilization with the disrupted ends approximated.

until the edema and tenderness decrease. This is an important note for patients who continuously stand at a plant or manufacturing work location. Light duty may be implemented if medium to heavy work is the worker's customary job. The employee/patient physical exertion levels are monitored.

Passive stretching should be discouraged when there is significant hemorrhage and edema. Ambulation (for lower extremity strains) or use of the injured muscle (in the upper extremity) should be gradual, with the patient working within his or her pain level. Progressive active exercises can be initiated to the point of pain. This phase of treatment can be accompanied by modalities for pain.

Third-degree strains may be immobilized in a splint, after which ice is applied and the extremity is elevated. The patient may be a candidate for surgical repair. Partial tears of the Achilles tendon and surgical tendon repairs may result in several weeks (i.e., 4–6) of casting. Immobilization may last for weeks, depending on the healing process of the tendon involved and the extent of the injury to the tendon, with modifications made during the course of the immobilization.

STAFFING

The staffing of on-site (at plant or company) or off-site (at medical clinic) operations is dependent on the size of the workforce. As the program's patient visits grow, staffing is adjusted. Staffing must cover the varied shifts of both hourly workers and salaried personnel. Plants may operate two production shifts per day with a maintenance crew for all three shifts, 7 days per week. There should be a fully equipped and staffed medical department either in the plant or off site to address employee health needs. The staff is accountable to the plant medical director, the plant personnel director, the plant manager, and the program administrator of the direct company contract supplier. Patients should drive no more than 10 miles to an off-site medical facility for workers' compensation care. This distance has been proven to keep employees close to work for ease of getting back to the job, thereby reducing lost work time, worker-replacement costs, and employer lost productivity. In the event of a life- or limb-threatening or emergency disease or illness, an ambulance usually stabilizes or transports the patient to the closest emergency department. Patients with urgent conditions that are not life- or limb-threatening may be treated on site. If on-site medical providers are not available, the worker may be taxied or a worker's supervisor may drive the patient/employee to the occupational medicine clinic and/or appropriate hospital/ambulatory care facility.

The occupational medicine physician working at the on-site facility is trained and knowledgeable about job physical demands and requirements. A registered, board-certified occupational therapist may be prescribed by the physician to perform an on-site job assessment or functional capacity evaluation of the patient. The return-to-work program is dependent upon an established disability assessment and plan to transfer the patient/employee to full duty or—based on capability and skills—to another job placement.

Medical staff participate in case management or ''grand rounds.'' Reevaluation and follow-up appointments continue until the case is closed. The employee is no longer a patient when the condition has resolved, permanent placement has been assigned, or the employee has returned to the same job at a preinjury/illness level. Staff identifies potential job compatibility through medical and workforce profiles. Temporary (usually lighter-duty) job placement may be found for a patient who is not determined to be permanently disabled. Thus, the number of open claims or cases at a company should be considered when setting up an occupational health program. The author advises reviewing the plant or company's OSHA logs 3 years back to familiarize the on-site team with the most frequently seen injuries and high-risk areas in need of prevention and ergonomic evaluation.

The number of occupational medicine physicians may include four that are board-certified or two boarded and two board-eligible for companies with 20,000 employees. The number of physician assistants may be varied based on coverage needed on all three shifts. The number of occupational health nurses is usually three for each day and afternoon shift. One nurse may work the third shift when production is down. Two physical therapists, two occupational therapists, a rehabilitation technician, and a secretary/biller for a 5-day-a-week rehabilitation operation from 7:30 a.m. to 6:00 p.m. is considered standard staffing. Fitness center staffing includes two certified athletic trainers from 5:30 a.m. to 10:00 p.m.

What percentage of the time is an M.D. present when a physician assistant or nurse practitioner is seeing patients?

Physician assistants (PAs) are independent practitioners who provide patient care services under the supervision and responsibility of a doctor of medicine or osteopathic medicine (100% of the time). Certified nurse practitioners provide supervised medical care as part of a team. See Appendix 4 for the nonphysician practitioner guidelines, which describes credentialing, supervision, and privileging. This appendix describes a physician assistant's treatment statutory role and the role of medical doctors in charge of the physician assistant.

Is an M.D. available by phone 100% of the time when a physician assistant or nurse practitioner is seeing patients?

A medical doctor is available by telephone 100% of the time when a physician assistant or nurse practitioner is seeing patients. The medical doctor is also available by beeper 24 hours a day. The physician assistant does not initiate treatment based on his or her reading of x-ray films before they are read by the medical doctor. The types of cases that are seen by the physician assistant or nurse practitioner versus the physician are common health problems such as infections and sprain/strain injuries, diabetes, high blood pressure, and consultations for preventive health and wellness. The physician assistant makes an appropriate record of findings during examinations, which is to be reviewed and countersigned by the supervising physician.

OSHA is designed "to assure so far as possible every working man and woman in the nation safe and healthful working conditions and to preserve our human resources." In administering OSHA, the Labor Department issues standards and rules for safe and healthful working conditions, tools, equipment, facilities, and processes. OSHA conducts workplace inspections to make sure that the standards are observed. It has been the author's experience that by performing a new-hire preplacement exam, a workers' compensation claim can be avoided.

Preplacement or new-hire evaluations have two purposes. The first is to determine whether the person has a health condition that may be aggravated by conditions in the workplace (e.g., a hernia). The second is to determine whether the person has a health condition that may place the safety or health of others at risk (e.g., drug abuse). A preplacement evaluation can include an evaluation of the musculoskeletal system in relation to the physical demands of specific activities of the job; a history and physical are provided. Evaluations are job-specific. For example, in a preplacement exam, a nurse whose job is working in an operating room will be given a latex-sensitivity assessment. Drug screening during a Department of Transportation (DOT) assessment is part of a ''car carrier'' truck driver's examination. A hearing conservation program may be administered to a police cadet who will spend time on a firing range. Visual acuity, pulmonary function, and respiratory fitness testing may be part of a preplacement evaluation for an hourly worker in an automobile body service department whose job it is to hold a spray gun to paint and repair a vehicle. An infection-control screening and immunization program is mandatory for health care workers involved in direct patient care. Workers in a hemodialysis unit will be offered a hepatitis B vaccination program.

The DOT driver regulations include drug testing performed according to DOT drug and alcohol testing procedures, which set forth the procedures for drug testing in the Federal Highway Administration (FHA) industries. All drug test results are reviewed by a physician medical review officer (MRO) before the results are reported to the employer. The occupational medicine physician and/or MRO complies with all mandated federal regulations for labs, also coordinating drug screening for preplacement and fitness for duty to ensure adherence to the National Institute of Drug Administration (NIDA) protocols. Drug testing is a two-stage process. First a screening test is performed. If the urine specimen is positive for one or more of the drugs, then a confirmation test is performed for each identified drug using state-of-the-art gas chromatography/mass spectrometry (GC/MS) analysis. GC/MS confirmation ensures that over-the-counter medications are not reported as a positive result. If the laboratory result is reported as positive, the physician MRO contacts the employee (usually by telephone or in person) to determine by interview if there is an alternative medical explanation for the drugs found in the urine specimen.

Preplacement and return-to-work exams require that occupational health providers perform all physical examinations as identified. The physician should provide information regarding the experience and/or special training identified relative to employment-based physical examinations. For job-related examinations and occupational health examinations, the physician shall evaluate the individual only on his or her ability to perform the essential functions of the job without endangerment to self or others. Employability shall be determined in accordance with the Americans with Disabilities Act (ADA). The physician shall advise the human resources department of restrictions, if any, or if a direct threat to health or safety exists. The occupational health providers have extensive experience in administering appropriate job-specific preplacement examinations consisting of a history and physical and surveillance, such as:

1. Examination for musculoskeletal impairments
2. Recommendations requiring heavy lifting of weights of 40–60 pounds
3. Blood lead testing
4. Alcohol and drug testing for commercial drivers (as related to DOT and the Federal Highway Administration)
5. Latex-sensitivity screening (lab: latex-specific IGE); hearing (audiogram); vision testing (titmus, laser exam, tonometry); vaccination (hepatitis B, three doses; tetanus)
6. Respirator exam, spirometry, respiratory fit testing
7. Lab antibody titers (rubella, rubeola, mumps, varicella, hepatitis B)
8. Lab tests (i.e., thyroid, biochemical profile, occult blood)

Unnecessary exposure of employees to ionizing radiation should be avoided. Regular or routine x-ray examinations of apparently well employees are to be discouraged unless required for special occupational exposures or to meet regulatory requirements. Chest x-rays should be performed on potential new employees at the discretion of the examining physician (e.g., prior occupational exposure to known toxic dusts is usually a reason to obtain a chest x-ray). Periodic chest x-rays should be provided for specific occupational risk populations as determined by the examining physicians if required by law or if clinically indicated (e.g., a ''B'' reading to evaluate asbestos esposure). Lumbar spine x-rays are not to be performed as a routine screening procedure for back problems. Where there are specific medical indications for such, a written order by the requesting physician is required. The on-site team uses guidelines in conjunction with workplace job duty assessments and the patient/employee's assessment of the job's physical demands.

Physicians are familiar with the applicable provisions of the ADA. The occupational medicine team is experienced in resolving conflicts between the mandates of the ADA and other laws as well as with inconsistencies related to distinctions in insurance coverage. The team has experience with making reasonable accommodations. Usually a company will have set guidelines for what is reasonable based on whether an intervention is too costly if an undue hardship for the company is involved, or if it changes the job process. Physicians and the on-site team recognize that observance of the ADA is an ongoing obligation.

APPENDIX 1: SAMPLE OF AN AUTOMOBILE COMPANY'S REQUEST FOR PLANT ON-SITE OCCUPATIONAL MEDICAL PHYSICIAN SERVICES

Introduction

Automobile Company is a major motor vehicle manufacturer, employing approximately 172,000 employees distributed among 136 locations.

As a major industry, the automobile company has an occupational health department, organized into a corporate occupational medicine department, which is responsible for workers' compensation care, policy, and medical management. Within the automobile company, there are several subdivisions such as health and safety, ergonomics, and employee assistance programs. Our interest here is for you to provide services to support physician staffing at our Metro-Detroit, Michigan, occupational medical clinics. The position of medical director for occupational health services is also required to be staffed. Further definitions and criteria are listed in the attached document.

Our goals are:

To have all Metro-Detroit and medical director physicians be supplied, trained, managed, and placed on-site by one supplier
To improve the quality and efficiency
To provide one source for professional credentialing services

Automobile Company will administer the contract services through its international health services operating department.

The plant occupational health services provided are:

1. First-aid and emergency care for all employees, contract workers, and visitors
2. Assessment and care for occupational illness and injury
3. Evaluation and referral for specialty and emergency (life or limb) occupational illness or injury care, in accordance with workers' compensation rules and regulations
4. Supportive care and/or triage for nonoccupational medical conditions to primary care provider

5. Medical examinations including preplacement, transitional work, fitness for duty, return to work, termination, and permanent disability
6. Medical examinations or laboratory or ancillary testing for monitoring of occupational exposures
7. Medical case management of workers' compensation illness or injury claimants
8. Medical case management of employees on extended (nonoccupational) sickness and accident disability leave
9. Interface with plant health and safety, industrial hygiene, ergonomics, and plant management

Automobile Company desires that interested occupational medical service providers submit a detailed quote that addresses the auto company's needs as specified in this request for proposal (RFP).

Section I: Scope of Services

First Aid and Emergency Medicine

Automobile Company's plants currently provide on-site first aid/emergency care by having registered nurses, licensed practical nurses, and physician assistants on site during each major work shift. At large plants, an occupational medicine physician is usually on duty during regular daytime shifts. First aid equipment and supplies are available at the work site. At some plants, a second and third shift registered nurse and physician assistant are on duty. Most occupational medicine physicians vary hours to cover all three shifts.

The supplier must provide emergency medical services at the plant or provide appropriate referral or transfer for all emergency cases to a local hospital emergency department.

The supplier must provide education and training for appropriate first aid of designated automobile company personnel, who may be called upon to assist in these emergency situations.

Occupational and Nonoccupational Illness and Injury

Automobile Company wants its employees/workers with occupationally caused illness or injury to receive prompt, expert medical care.

The supplier shall provide medical care for employees with occupationally caused illness or injury. Referrals can be made to employee/patients requiring highly specialized facilities or treatments. Lost work time will be monitored even when employee/patient is treated off site by a specialist.

The supplier will treat or refer, and follow, all occupational illness or injury cases, to facilitate appropriate care and placement in accordance with Automobile Company policies and applicable workers' compensation benefit programs.

All supplier personnel shall have a thorough knowledge of the workers' compensation rules and regulations, applicable to the state and federal standards. The supplier shall effectively work with plant compensation and benefits personnel and with other workers' compensation agencies and departments who are acting as agents for Automobile Company.

The goal of the on-site occupational health provider is to care for work-related illness or injury, but one may encounter a nonoccupational injury/illness that must be medically stabilized and initially treated. In any case the Automobile Company medical units are occupational health clinics on site. All conditions are reviewed and assessed.

Medical Examinations: Preplacement, Fitness for Duty, Return to Work, and Termination

Automobile Company requires preplacement examinations for all "regular" (greater than 120 days duration) new hires. Summer interns or vacation replacement personnel are required to fill out a history and physical self-assessment form, pass a five-panel drug screening of the urine, and complete a musculoskeletal and new hire physical exam. Each new worker's success in the hiring process is contingent on a negative drug screen. There is a contracted laboratory for this service. All drug test results will be reviewed and interpreted by a physician medical review officer (MRO) before results are reported to the company contact. A chain of custody will ensure that the specimen's security, proper identification, and integrity are not compromised. Drug testing will be performed according to the DOT drug and alcohol testing procedures and rules in the FHA industries. Confidentiality is maintained. All those preplacement candidates who will, upon a negative drug screen result, be hired on a "regular" basis shall have a musculoskeletal and physical examination to evaluate their ability to perform their proposed assigned job duties with respect to ADA law.

Occupational health physical examinations for return-to-work, transitional work, medical fitness (e.g., crane operators), transfer, or termination examinations may be performed. The examinations noted require that an appropriate medical and occupational history and a physical examination be completed. Audiometric examination, chest x-ray, visual acuity, electrocardiograms, or blood or urine tests (except "dipsticks") may also be completed as related to the job description, title, or duties and/or as applicable by law.

The on-site supplier of health professionals will be expected to review the examination results to determine the individual's current health status. The on-site physicians will be responsible for documenting work restrictions and entry of those restrictions in the computerized information systems.

Hours of service, the volume and type of examination will vary depending on patient and plant. Response time to plant needs is imperative and may call for extended daily physician work hours as well as weekend or holiday coverage to meet plant production schedules.

Medical Monitoring and Tests for Occupational Exposures

Some Automobile Company plant employees are occasionally exposed to substances requiring periodic medical or surveillance monitoring. Employees who may have past asbestos exposure are examined on a periodic basis.

At some plants, significant numbers of employees may be exposed to noise levels and may require occupational hearing loss protection that includes an annual audiometric test. Also, some plants may have respiratory protection programs that require periodic medical evaluation or respirator users and pulmonary function testing.

It is the responsibility of the on-site occupational medicine physicians working with the plant to identify employees who require medical evaluations, monitoring, or specific tests in compliance with OSHA regulations. The industrial hygiene department and health and safety at the plant are responsible for personal protective equipment and training/educational programs for their employees. The supplier shall work cooperatively with health and safety to provide the requested medical services and to update practices in compliance with the laws governing workers' protection.

The plant occupational medicine physician supplied on site will be familiar with the plant work environment, job duties, job descriptions, and production processes. The plant physician will tour the operation at least monthly or daily, if necessary, to make appropriate recommendations for job modifications, ergonomics, worker activity, and/or toxicology evaluation.

The plant occupational medicine physician shall be responsible for presenting case review at case management meetings and shall participate in ergonomics, disability management, absentee, and light duty programs. The physician shall assist the company in developing a strategy for implementation and compliance with OSHA and in recordkeeping.

After the first 10 days in some states (e.g., Michigan) following an injury or illness, a workers' compensation patient has a right to choose his or her own physician for care (State of Michigan, Workers' Compensation Act, 418.315, Section 315). A company-preferred provider list may be utilized for laboratory services, hospital MRI, or rehabilitation, developed with the third-party administrator (TPA).

Medical Evaluation of Employees with Workers' Compensation Claims

The supplier shall conduct case management of employees with workers' compensation claims. A medical evaluation of a patient/employee on a 2-week basis to help determine medical status, rehabilitation progress, and/or medical restrictions upon return to work will be completed. Such evaluations may include a review of medical information and examination of the specialized opinions and communication with the specialty physicians. IMEs may be ordered and utilized as second opinions. Computerization of medical records will assist in case management and settlement of disputed claims.

Medical Evaluation of Workers on Nonoccupational Disability Cases

Nonoccupational sickness and accident cases may be reviewed to assist in case management and job placement.

Records, Reports, and Forms

Automobile Company and the on-site occupational medicine physician shall provide records, reports, and forms regarding clinical guidelines, treatment protocols, and best-in-class services. Data shall be automated in the information systems for outcome measurement and monthly summary reports of lost time injury/illness.

The supplier shall be familiar with OSHA requirements for records and reports and shall be responsible for compliance with these requirements.

The supplier shall maintain medical records and personal medical information in accordance with policy and in conformance with medical record confidentiality and medical ethics.

Medical records and reports shall be legible, identifiable, and submitted on time and according to the medical records standards and policy and procedures governing the protection of these records.

Immunizations

The on-site physician shall follow the recommendations of the U.S. Department of Health and Human Services–Centers for Disease Control as related to bloodborne pathogens standards, administration, compliance, and consultation. Vaccinations (hepatitis B) and immunizations will be offered to high-risk patients. The physician will advise the patient of necessary overseas recommendations of immunizations and related overseas travel.

Epidemiologic Surveillance

Supplier shall analyze high-risk or rapid widespread of a disease. Research and/or investigation must be reported to plant management and international health services operations departments. Deaths must also be reported immediately to both the plant management and health services department.

Health, Education, Health Promotion, and Wellness

Supplier shall provide educational materials on smoking cessation, diabetes, blood pressure, and healthy living. Health promotional wellness fairs shall be held in the fall of each year. Employees/patients shall be encouraged to attend for blood pressure checks, glucose testing, and body fat composition analysis, among other screenings (i.e., vision, cholesterol, and prostate-specific antigen). The company rehabilitation and fitness center usage shall be monitored. Employee assistance programs will be recommended as needed.

Section II: Staffing and Personnel

Currently, there is a need for 12 plant physicians; one full-time medical director will be on site to oversee the other 11 physicians working at various plants. The following is the physician coverage needed:

Plant	Hours
Transmission	40
Tech Center	80
Axle	40
Assembly Plant 1	40
Engine Plant 1	20
Glass	20
Engine Plant 2	40
Assembly Plant 2	40
Stamping Plant 1	40
Assembly Plant 3	40
Stamping Plant 2	40
Truck Plant	40

The supplier will provide board-certified or board-eligible occupational medicine physicians with plant experience in the health services described in this request for proposal (RFP). Auto company considers staffing to be the key factor to high-quality program implementation. Physicians shall be recruited and employed by the supplier. The supplier shall be responsible for the compensation, benefits, and professional liability coverage for its physicians. Replacement workers shall be provided during vacations and/or illness.

Medical Director—Job Description/Duties

The medical director shall be board certified in occupational and environmental medicine and shall be licensed to practice medicine in the state. Clinical responsibilities shall be the same as a plant physician, described below with administrative, leadership, policy and procedure, educational implementation, execution, and delivery. This position shall be housed at the technology center. The medical director's responsibilities will typically split 50% clinical and 50% administrative. The supplier shall provide a back-up plant physician for times when the medical director's duties require absence from clinical work.

The medical director's duties are as follows:

1. Implement with plant physicians on policy, education, and delivery-of-care programs
2. Attend corporate staff meetings representing medical
3. Review medical cases, consult, close complex cases
4. Review preemployment examinations and workers' compensation care plans
5. Consult with the corporate legal department on cases requiring independent medical evaluation, settlement, or total and permanent disability
6. Ensure that services provided are consistent with good clinical care and contract obligations
7. Participate in evaluation and selection of continuous quality improvement processes
8. Participate in regulatory compliance and streamline procedures for treating clinicians
9. Interface with plant management and medical
10. Report to both on-site auto company plant management and administrative director of hospital

Plant Physician—Job Description/Duties

Plant physicians shall be licensed to practice medicine in the state and will be board certified or board eligible in occupational medicine. Two years' experience in occupational medicine is preferred.

Each plant physician is the manager of his or her respective plant medical department with supervisory duties and responsibilities for on-site nursing. Medical management of patient cases is the responsibility of the physician. The plant physician will communicate with plant management daily.

Plant physicians will work jointly with ergonomics, human resources, job placement, plant management, and health and safety.

The supplier will personally interview each physician candidate and submit a curriculum vitae to the plant prior to the plant's formal interview of each potential candidate. Credentialing status must be approved by the supplier.

Each plant physician has clinical responsibility for employee/patient health at the plant. His or her knowledge of the working environment, exposures, prevention, case management, and ergonomics is imperative. He or she will complete physical exams, preplacement assessment to determine work activity, restrictions, light duty, transitional work, and placement. Additional responsibilities include assisting in the implementation of compliance with federal and state regulatory standards.

Each plant physician will be guided in his or her professional conduct and will adhere to high standards of medical ethics (e.g., American College of Occupational Medicine Code of Ethics). In addition, the plant physician is responsible for compliance to all pertinent state and federal laws governing the practice of medicine.

Plant physicians will monitor preventive maintenance for existing and all future medical equipment. He or she will direct the staff to implement corrective action/repair as required. Documentation in logs and reports of such actions will be the responsibility of the plant physician and his or her staff.

Section III: General

The quote shall be expressed on a dollar-per-hour basis or as a monthly management fee for the following plant provider: (a) medical director or (b) plant physicians. The supplier shall be provided with a purchase order for scope of work, and hours worked.

The supplier is required to comply with rules and applicable policies and procedures. Both the hospital administrative director and plant manager will evaluate the physician performance at least annually.

The supplier shall provide physician backup for sick days, vacation days, and continuing education days or meetings.

The supplier shall be responsible for professional continuing education, licenses, associations, malpractice insurance, other respective liability coverage and ongoing professional training of its staff. Inservices at the on-site plant shall be attended by the physician.

Occupational medicine physicians are employees of the supplier.

Staffing changes (personnel, scheduling, hours, etc.) are subject to approval by the buyer of services. Production schedules may effect the medical staff hours of work.

The physicians will remain current in all areas of occupational and environmental medicine in order to supply state of the art care and consulting services.

The supplier will be responsible for appropriate staff training for all regulatory compliance items (e.g., bloodborne pathogen standard, respiratory fit, immunizations, etc.).

Current copies of licenses, certifications, and credentials must be on file at the plant and supplier location.

Supplies—the medical unit will order medical supplies from approved suppliers. This will not be the responsibility of the supplier. Recommendations will be made for the appropriate stock and inventory of supplies on site.

Section IV: Program Administration and Oversight

The supplier, under the direction of the administrative director, shall perform an annual employee performance review. Input for the review will be gathered from plant management, the international health services operations department as well as the designated medical director. A formal quality assurance program shall be provided quarterly along with continuous quality improvement reports, ''best in class'' approaches, customer satisfaction reports, and outcome studies of quality and cost.

Physician standing orders, policies and procedures, and quality practices will be provided on site. Physician and nursing directives for treatment services signed and dated by the physician and nurse assigned to the respective location will be current. A current copy of this must be placed on file in each plant medical department. Policy, procedure, and ISO-9000 quality manuals will be on site, and both supplier and buyer written manuals will be followed. For the delivery of patient care practices, the suppliers' clinical manuals will take precedence.

Section V: On-Site Environment Facilities, Equipment, Supplies, Office Maintenance/Utilities

Facilities

The buyer will provide adequate partitioning. The size and configuration of the plant medical departments vary from plant to plant, but generally contain reception, examination, treatment, and rest rooms and clean and soiled storage areas. The buyer will provide utilities, phones, faxes, and computers.

Equipment

Equipment at existing plant medical units will be utilized (i.e., electrocardiogram, x-ray units). For consistency and cost-effectiveness, new equipment shall be proposed to plant purchasing. It is the physician's responsibility to assure that the staff keeps all equipment calibrated and functioning.

Supplies

The plant physician is responsible for obtaining and maintaining all necessary registrations, licenses, etc. associated with ordering both prescription and nonprescription and controlled and noncontrolled items for their staff.

Guidelines for the dispensing of drugs and medications to employees/patients will be adhered to by medical staff.

Where the use of company letterhead, stationery, or its equivalent is appropriate, Automobile Company will furnish same in accordance with usual business operations.

Laundry, oxygen, audiometric calibration/repair, and x-ray support/maintenance services are the responsibility of each respective plant medical unit. Annual leases to secure services as needed will be obtained.

Offices

Each plant medical unit has a physician office for the physician for patient confidentiality.

Maintenance/Utilities

Automobile Company shall provide customary ''landlord'' services (i.e., utilities, computers, telephone services, basic housekeeping, general maintenance, security services, and in-plant transportation). Automobile Company will pay the cost of maintaining equipment.

Section VI: Supplier Qualifications and Experience

Responses to this request shall include information as to the supplier's past ability to direct-contract on site for occupational health programs and qualifications, plus prior experience in performing said services on site in plants. The supplier's ability to manage, execute, and operationalize and the supplier's expertise in occupational health will be considered.

Section VII: Timing/Delivery

Automobile Company needs to select, cost out, and interview suppliers who will be able to perform the requested work based on their submission and an in-depth formal presentation of their proposal. Responses to this request must be submitted on or before (date).

Section VIII: Presentation and Submission of the RFP/Quote

Your technical proposal shall at a minimum contain the information specified below in the following format:

- **I.** Understanding the scope of the work and practice expertise
- **II.** Innovative technical approach
- **III.** Medical management
  - **A.** Organizational plan
  - **B.** Management approach and staffing levels

IV. Organizations qualifications and prior experience in plant settings
V. Physician respective candidates and curriculum vitae (C.V.)
  A. Medical director
  B. Plant physicians
VI. Expound on physicians plant occupational medicine experience
VII. Hourly rate (each physician shall be paid on a straight time basis)—overtime rates are not permitted.

Please submit five (5) originals of your response proposal by (date). Deliver to buyer Mr. Brown at World Headquarters, Automobile Company, Executive Drive, Detroit, MI.

APPENDIX 2: OCCUPATIONAL HEALTH CLASSIFICATION DESCRIPTION

Job Title: Occupational Health Nurse

Date:

Job Summary: (Includes responsibilities)

Under direct supervision of the director, occupational medicine, assists in all phases of occupational health care. Evaluates and treats work-related illness and injuries, conducts new hire physical exams, assists with health screenings and immunization program.

Duties (brief summary only):

1. Assesses and evaluates all injuries/illnesses and medical complaints thoroughly and refers individual for appropriate medical treatment.
2. Recognizes, prioritizes, and responds to medical emergencies promptly and efficiently.
3. Treats and medicates according to doctor's directives, policies, and procedures.
4. Provides education, explanation, and instructions to patients about condition and treatment.
5. Develops and implements a nursing care plan that provides for continued care and treatment, rehabilitation, and return to work.
6. Ensures that health records are maintained in compliance with OSHA and state and federal regulations.
7. Releases medical information only to authorized personnel by appropriate procedures. Maintains confidentiality of medical records.
8. Reviews physical exams, special exams, and physical assessments for accuracy and completeness.
9. Counsels and/or instructs troubled employees as necessary.
10. Ensures follow-up of special problematic work-related injuries and illness.

11. Implements and coordinates existing and newly introduced programs in a timely manner.
12. Is familiar with company policies, including workers' compensation and OSHA regulations and uses them as guidelines when dealing with employees.
13. Seeks guidance from supervisors when necessary.
14. Provides education, support, and motivation in the areas of health and safety.
15. Evaluates health programs and nursing services periodically and revises them to meet changing needs.
16. Records all pertinent medical and legal information legibly and accurately on proper forms. Ensures forms are submitted in a timely manner and filed as required. Inputs data into the information systems.
17. Ensures medical equipment, supplies, and drugs are properly maintained.
18. Ensures that exam rooms are supplied and work areas are kept clean.
19. Anticipates use of emergency equipment and is knowledgeable as to its use.
20. Provides thorough on-the-job training for staff as necessary.
21. Performs other related duties as necessary.

Supervision

Reports to the physician and administrative director, occupational health and industrial rehabilitation department, or designated representative. Does not supervise any other employee.

Minimal Qualifications

1. Registered nurse with current state licensure required.
2. Minimum of 5 years' experience or certification in occupational health nursing required.
3. Demonstrated clinical knowledge or occupational health nursing and the analytical ability necessary to formulate effective nursing care plans.
4. Demonstrated ability necessary to formulate effective nursing care plans.
5. Familiarity with basic work processing programs is preferred.
6. Familiarity/working knowledge of OSHA, ISO-9000, and other federal and state occupational health requirements/regulations.

This description is intended to indicate the kinds of tasks and levels of work difficulty that will be required of positions that will be given this title and shall not be construed as declaring what specific duties and responsibilities of any particular position shall be. It is not intended to limit or in any way modify the right of any supervisor to assign, direct, and control the work of employees under his/her supervision. The use of a particular expression or illustration describing duties shall not be held to exclude other duties not mentioned that are of a similar level of difficulty.

APPENDIX 3: DESCRIPTION OF PHYSICIAN ASSISTANT DUTIES

1. Is the agent of the supervising physician, even to the extent that patient confidentiality holds with the physician assistant just as with the physician.
2. May not assume the "ultimate responsibility for the quality of medical care services."
3. May prescribe drugs other than controlled substances. The PA's name and the supervising physician's name should be listed on the prescription.
4. At a minimum, the physician assistant is competent to:
   a. Do the initial follow-up evaluation of patients of various age groups in any setting to elicit a detailed and accurate history, perform an appropriate physical examination, and record and present pertinent data, including interpretive recommendations, in a manner meaningful to the physician.
   b. Perform or assist in the performance of routine laboratory and related studies as appropriate for a specific practice setting, such as blood studies, urinalysis, and electrocardiographic tracings.
   c. Perform routine therapeutic procedures such as injections, immunizations, and the assessment, suturing, and care of wounds.
   d. Instruct and counsel patients regarding physical and mental health, including matters such as nutrition, illness, treatment, or lifestyle risk factors.
   e. Perform the following functions on site at the company: patient assessments and treatment, making ergonomic inspection of plant, recording patient progress notes, accurately and appropriately transcribing or executing standing orders and other specific orders at the direction of the supervising physician, and compiling and recording detailed progress reports and narrative case summaries.
   f. Deliver or assist in the delivery of services, including the review and monitoring of treatment and therapy plans, to patients requiring initial or continuing care at the occupational medicine clinic.

   g. Execute and treat life-threatening emergency situations.
   h. Interact with those community health services and other community resources that will facilitate the patient's care and continuity of care.
5. May alternate with a physician for required patient visits on second and third shift at the worksite clinic.
6. May perform routine care at the on-site clinic.
7. May order and/or take specimens for the testing of drugs, hepatitis B, and make the appropriate documentation of same.
8. May make calls or go on rounds in the plant, emergency vehicles, ambulatory care clinics, hospitals, intermediate or extended care facilities, health maintenance organizations, nursing homes, or other health care facilities to the extent permitted by the policy, law, rules, or regulations of the facilities or organizations under the supervision of a physician.
9. May assume the delegated responsibility from the supervising physician of performance of medical care services if the delegation is consistent with the physician assistant's training.
10. May perform routine visual screening or testing, post-op care, or assistance in the medical care of diseases of the eye under the supervision of a physician.

A physician assistant shall not:

1. Undertake or represent that (s)he is qualified to assume medical care that (s)he knows or reasonably should know to be outside his/her competency. Such undertaking or representation is prohibited by law.
2. Perform any tests to determine the refractive state of a human eye or to treat refractive anomalies of the eye.

APPENDIX 4: DUTIES OF PHYSICIAN IN CHARGE OF A PHYSICIAN ASSISTANT

1. Must verify the PA's credentials.
2. Must evaluate the PA's performance.
3. Must monitor the PA's practice and provision of medical care services.
4. Is responsible for the clinical supervision of each PA to whom the physician delegates medical services.
5. Must maintain a permanent, written record indicating the physician's name and license number, together with a list of each PA the physician is supervising and their license numbers.
6. On site at a company or plant may not delegate a task when the regulations specify that the physician perform it personally or when the delegation is prohibited by state law or by the facility's own policies.
7. This description is intended to indicate the kinds of tasks and levels of work difficulty that will be required of positions given this title of physician assistant and shall not be construed as declaring what the specific duties and responsibilities of any particular position shall be.

2

The Benefits of the On-Site Occupational Health and Rehabilitation Model to the Whole Organization: Management, Employee/Patient, and Union

In this chapter the author describes how her model will assist the company in managing occupational health services for its employees. The model describes processes that rely on the resources of a large, integrated health care delivery system to provide workers' compensation care at plants, on site. Interest in aggressive management of work-related injuries and illness is not new. The model works best with business units that are organized into an occupational health product line to serve the needs of customers, including claims processing, specialty and ancillary services, routine and emergency care, and data and information systems.

The on-site model responds to the needs of employers/company's employees, workers' compensation representatives, and third-party payers and is designed to reduce both health care and indemnity costs. It also allows the company to take advantage of one-stop shopping for clinical and laboratory services and to access multiple outpatient and inpatient sites for after-hours and emergency care. The range of specialized and support services available to the company through this relationship will provide high-quality, low-cost care. The best approach includes linking company medical and case-management staff to the occupational health practice, budget, and quality initiatives. The key service profile is described in the following section.

PROGRAM ATTRIBUTES AND SERVICES PROVIDED

On-site facilities are identified and physicians who have qualifications in the delivery of workers' compensation care are available to offer employees timely access to care. Medical services, including injury treatment for company employees, will be provided at locations on site by experienced occupational medicine physicians who have successfully completed a rigorous credentialing process. Physicians are expected to monitor an employee's condition directly at work. The role of the on-site physician includes evaluation, immediate treatment, and recommendations regarding further treatment. The disease-management strategies will be relied upon to avoid treatment patterns or levels of care that are inappropriate or excessive. Space is provided by the company for on-site medical services. The author has worked with employers to design and build the on-site clinic. In companies with no available space she has recommended a trailer and has provided staff and all equipment and supplies for the trailer clinic. On-site space has been as small as 500 square feet for a 200-employee company to 8000 square feet for a medical rehabilitation and fitness center on site in a manufacturing company employing 3800 workers.

The on-site team is able to routinely call on its own physician consultants, who are prominent in their specialties to provide expert advice or clinical services. Clinical specialists (e.g., neurosurgeons) can be available to assist the on-site physician by conducting patient evaluations, reviewing cases, and suggesting the most reasonable selection of diagnostic tests. The occupational and environmental medicine physician specialist on site validates treatment plans and helps to maintain compliance with regulations of the Occupational Safety and Health Administration (OSHA). The board-certified occupational and environmental medicine provider makes the final decision regarding the employee's ability to return to work.

Recognizing that the company desires to explore the effectiveness and feasibility of utilizing occupational medicine physician services on site at facilities with high numbers of OSHA recordables, the length of these physician contracts could range from several months to one year or be ongoing with the company having an option to renew. The use of physician services must improve the quality, efficiency, and cost-effectiveness of occupational health services for employees.

The *scope of services* provided by on-site company board-certified physicians includes:

1. First aid and emergency care for employees, contract workers, and visitors:
   Occupational injuries and illness determination, evaluation, and treatment
   Emergency medical care
   Emergency cardiac life support (ACLS)
   Emergency trauma management
2. Limited definitive care for minor occupational illness and injury:
   Work-related sickness and accident care determine work activity, restrictions, light duty, transitional work, and placement
   Routine radiology
   Spirometry
   Exposures (toxic, bloodborne, allergic, hazardous materials or conditions)
   Case management
   Ergonomics
3. Assessment and referral for specialties (e.g., neurosurgeon), occupational illness, or injury care in accordance with workers' compensation and regulations:
   OSHA-200 and 101 log forms and assistance with the Americans with Disability Act (ADA)
   Department of Transportation (DOT) and alcohol testing
   Evaluate cost and quality of outside medical and wellness services
   Routine and random screening to complete preplacement/new hire assessment
   Workers' compensation claims evaluation and longitudinal case management along with specialty providers
4. Limited supportive care and referral of *nonoccupational* medical conditions:
   Injury/Illness assessment
   Appropriate coordination of care with primary/family practice physician
   Emergency trauma management and medical care
   Some routine illness treatment and referral for more complex problems
5. Medical examinations (e.g., preplacement/new hire, transfer, fitness, return to work, and termination):
   Preplacement/new hire evaluations (i.e., history and physical, musculoskeletal evaluation, and physician exertion levels)
   Employee total and permanent disability determination and application
   Absentee control programs
   Disability management programs

Return-to-work programs

6. Medical examinations or tests for monitoring of occupational exposures:
   Medical surveillance physical examinations
   Bloodborne pathogens and hazardous materials
   Regulatory waste training
   Hearing conservation program and audiometrics testing
   Ergonomics program
   Vision testing
   Spirometry
   Electrocardiogram
7. Medical status evaluations to assist in the case management of compensable illness or injury claimants:
   Workers' compensation determination
   Return to work following workers' compensation sickness or accident
   Longitudinal case management
   Rehabilitation and work conditioning, on-site job assessment, and functional capacity evaluation referral and follow-up
   Safety committee and collective bargaining agreement participation
8. Medical status evaluations to assist in the case management of employees on extended (nonoccupational) sickness disability leave:
   Return-to-work evaluations following nonoccupational sickness or accident
   Worksite analysis for ADA compliance and accommodation restrictions
9. Assistance and advisory services to plant safety, industrial hygiene, job placement, and management personnel:
   Computerized medical information data at company clinical sites for outcome, continuous quality improvement, "best-in-class" approaches, customer satisfaction, and quality and cost studies
   Regular in services (case conferences, education programs, etc.)
   Prevention, health education, and health fair programs
   Guidelines for dispensing of drugs and medications, both prescription and nonprescription and controlled and noncontrolled

Providers will treat or refer and follow all occupational illness or injury cases to facilitate appropriate care in accordance with company policies and applicable workers' compensation benefit programs. In order to supply companies with the highest-quality services at the lowest cost, medical staff will be qualified by virtue of appropriate experience and credentials to deliver the services on site. Table 1 shows examples of medical care provided. Occupational health providers will operate and manage the on-site facility with expertise, appropriate staffing

Table 1 Medical Care Provided On Site and Not Requiring Specialized Facilities or Treatments

Routine treatment	Physical evaluations	Prevention
Lacerations	Preplacement/new hire evaluations	Routine vaccines:
Contusions	Sick leave/return to work	Tetanus
Strains and sprains	X-rays	Influenza[a]
Some fractures	Pulmonary function	Varicella[a]
Burns	Audiometry	Pneumococcal[a]
Eye injuries	Internal periodic exams:	Hepatitis[a]
Exposures and poisoning	Power equipment operators	*Lifestyle programs*[a]
Skin diseases	Crane operators	Smoking cessation
Respiratory conditions	Respirator users	Health risk appraisals
Repetitive trauma	Exposure evaluations	Stress reduction
Common illnesses (cold, headache, etc.)	Regulatory screening (drivers)	AIDS education
	For-cause and random screening	Body fat composition analysis and weight reduction program
		Blood pressure education
		Diabetes education

[a] Available upon request.
Source: Courtesy of Gregory Preston, M.D.

levels, and at shift hours agreeable to the company. The ordering of supplies and the maintenance of medical and office equipment is negotiated between the employer and the provider.

CARE OF ILLNESS AND INJURY—OCCUPATIONAL AND NONOCCUPATIONAL

Appropriateness of care for disability management involves the physician, who will be expected to deliver certain core services based on his or her occupational health training and to refer to specialty services when there is a likelihood that the patient will benefit from a specialist's care. The physician relies on disease-management strategies and clinical guidelines to assist providers with making the determination of when to perform expensive diagnostic tests, for the best practice approach to treatment of injuries, and for making timely referrals to subspecialists (i.e., workplace asthma protocol).

A major characteristic of the on-site model is that the primary occupational medicine physician and the subspecialist work together to determine the best treatment plan for an individual patient. They share the same system of medical records, information systems, laboratory and ancillary services so that the patient's care appears seamless and communication between specialist and referring occupational medicine physician is timely and supported by sufficient medical data. The on-site model proposes making the same information available to companies through its on-site medical information system and by careful coordination between company sites and referral sites through the central office of the department of occupational health. Any outside (off-site) medical services will be closely monitored and controlled by the on-site plant physician.

On-Site Policies and Procedures

The on-site model includes working with a number of workers' compensation (WC) agencies who are acting as agents for companies (i.e., the auto industry) and are familiar with their policies. The department of occupational health organizes multidisciplinary "grand rounds," which include a discussion of outstanding WC cases between medical and rehabilitation staff, benefits managers, union representatives, corporate human resources staff, etc. These grand rounds are invaluable in creating a team approach and familiarizing all participants in the elements affecting delivery of workers' compensation care. Grand rounds is a monthly meeting held with members of the medical-management process to discuss all open WC patient cases and may be also called a "case-management meeting." Patient status in occupational medicine and industrial physical therapy and occupational therapy are discussed. Suggestions can be made regarding continuation of therapy, changes in the treatment plan, and readiness for the patient to resume work with or without restrictions. This mechanism is used to ensure that patient cases are managed appropriately and that clinic visits are approved by the physician, workers' compensation representatives, and therapists. The clinical diagnosis and outcomes are reviewed to foster discussion, education, and rehabilitation goal attainment.

For example, early in my career at a steel company, I had been referred a patient/worker with a diagnosis of status—post–carpal tunnel release surgery. As the registered occupational therapist who specialized in hand treatment and the administrative director responsible for the success of the on-site occupational medicine and industrial rehabilitation clinic, I decided to look at the case. The patient had atrophy of his right dominant hand and forearm and I could visibly see the flexor tendons in his forearm. I could almost reach into the forearm and pull the tendons out, there was so much atrophy of the forearm. I did not believe that the patient had had carpal tunnel syndrome to begin with or that he should have had surgery. There appeared to be compression at the C-7–C-8, C-8–T-1 spinal levels. Grand rounds were scheduled for the next day, and I informed the on-site physician of my findings immediately. The on-site physician told me that he would see the patient right away. I suggested that the patient have an magnetic resonance imaging (MRI) examination, looking for perhaps a mass or cyst pressing on the cervical section of his spine causing nerve compression. The patient was an assembly line worker who could not lift objects heavier than 5 pounds with his right hand following the surgical intervention and was receiving 80% of his normal wages in workers' compensation benefits.

The on-site occupational medicine physician had never seen the patient before because this was literally the first week of operation for the on-site company clinic. The patient/worker had gone off site to a local physician, who had diagnosed him and then made a referral for carpal tunnel release surgery to a private practice surgeon. Once the on-site occupational medicine physician saw the patient, he was also convinced that the patient needed an MRI.

The next day at grand rounds both the on-site occupational medicine physician and I reported our observations. The steel company human resources, workers' compensation, and safety representatives, third-party administrator, plant management, and the union placement coordinator were also concerned with the outcome and findings of the MRI. The case was discussed in terms of an evaluation of anticipated date of recovery or need for total and permanent disability determination. The MRI was scheduled through an electronic referral system that allows occupational medicine physicians to get approval for specialist referral and appointment. This takes just minutes and reduces the process time for the patient to walk out of the on-site clinic with a referral to the specialist.

The results of the MRI were received within 24 hours by the on-site occupational medicine physician at the plant, the patient was found to have a tumor pressing on his spine causing the flexor carpi radialis, flexor pollicis longus, flexor digitorum superficialis tendon involvement with thumb metacarpal phalangeal flexion, and the thumb interphalangeal flexion weakness. The pronator teres muscle innervation was impaired along with the pronator quadratus. The tumor tuned out to be benign. The muscles innervated by the medial nerve in the forearm and hand were affected by the tumor.

The patient was grateful that the on-site team had appropriately diagnosed his condition. Since the plant had originally paid for the symptoms as carpal tunnel syndrome under the workers' compensation benefit, the plant agreed to pick up the cost for the surgery to remove the tumor. Had the on-site team been part of the company only months earlier, the patient's condition would have been diagnosed correctly. The condition was systemic in nature and it would have been denied as a workers' compensation claim and billed under the patient's Blue Cross plan.

The grand rounds team instructed the plant to settle the case for $132,000. The patient was 61 years old and unable to return to his original job. At his current annual salary and life expectancy, it was cheaper for the company in the long run to offer a one-time payout to the patient/employee. The patient accepted the settlement and expressed sincere appreciation for having received good medical attention and care by the on-site providers. The indemnity cost savings to the employer/company can be calculated at $656,800 if the patient/employee lived to be 78. Eighty percent of wages or $46,400/annually × 17 years (life expectancy of 78 years) = $788,800 in indemnity payments. Had the company not accepted the case as work related due to the systemic nature of the realized illness, costs would not have accrued in the areas of health care, mileage, productivity, reserves, indemnity, and legal fees to settle the case.

This case is one of many that have been discussed in grand rounds. Typically the cases involve patients who, based on ability, can work in light-duty or transitional work or until, through the course of physical and occupational therapy, they regain strength for repetitive work or are conditioned to return back to work.

Working with State Workers' Compensation Regulations

Occupational health providers should have a thorough working knowledge of workers' compensation regulations and procedures, have a reputation for expertise, and have managed care experience within the delivery of occupational medicine. Practitioners closely coordinate their activities with company WC representatives and benefits personnel to manage access to services and to reduce waste in the delivery of WC care to company employees. On-site physicians will also assist in the determination of compensability.

Nonoccupational Illness or Injuries

Minor nonoccupational injuries and illnesses that do not require special facilities or consume a great deal of clinic time may be treated on site, especially if it is clear that productivity will be maintained. Generally, patients who require continuing care for a medical problem that is not work related will be referred to their personal physician (e.g., for arthritis or diabetes). The plant physician or nurse may make a decision that a problem requires immediate attention or is a simple issue that can wait for routine care.

MEDICAL EXAMINATIONS: POSTOFFER PREEMPLOYMENT, FITNESS, RETURN-TO-WORK, DISABILITY, RETIREMENT, AND TERMINATION

Urine Drug Testing

On-site providers will coordinate the company drug-testing program. On-site providers will conform to the U.S. Department of Transportation urine specimen collection procedures. Direct nursing staff will be trained in the proper method for urine specimen collection, storage, and transportation. The medical review officer (MRO) will report initial negative results within 24 hours. Positive results with chromatography/mass spectrometry (GC/MS) confirmation will be reported to the designated company staffperson no later than 48 hours after specimen collection.

All confirmed positive test results are reviewed and interpreted before they are reported to the employer. If the laboratory reports a positive result, the MRO contacts the employee in person or by telephone to conduct an interview. If the employee can provide an alternative medical explanation for the positive urine test and provides acceptable documentation to the MRO, results judged to reflect legitimate medical use of the prohibited drug(s) will be reported to the employer as negative.

Drug Testing of Newly Hired Employees

Urine testing of all new hires from companies choosing to test include drug screening for the following five drugs and are conducted as part of the replacement evaluation:

1. Marijuana (tetrahydrocannabinol metabolites)
2. Cocaine
3. Amphetamines
4. Opiates (including heroin)
5. Phencyclidine (PCP)

Through experience with placing occupational health in the banking industry, on-site services include the use of a 10-panel drug screen. The handling of large quantities of currency by the worker has prompted 10-panel use. The drugs tested for include:

1. Amphetamine
2. Phencyclidine (PCP)
3. Cocaine
4. Opiates
5. Barbiturate
6. Benzodiazepine
7. Methadone
8. Methaqualone
9. Propoxyphene
10. Cannabinoids

Random Drug Testing

On-site providers may be responsible for conducting random, unannounced drug tests. The total number conducted each year must equal at least 50% of safety-sensitive drivers: some drivers may be tested more than once each year and some may not be tested at all, depending on the random selection. Random drug testing does not have to be conducted in proximity to the performing of safety-sensitive functions. However, once notified, a driver must proceed to an approved collection site for urine specimen collection. The Federal Highway Administration (FHA) periodically issues updated proposals that permit adjustment to random drug-testing procedures. On-site providers should assist companies in this process.

Record of Drug Test Results

Drug test results and records are kept under strict confidentiality by the employer/company, the testing laboratory, and the MRO. Results cannot be released to anyone other than the MRO without the written consent of the patient/employee. Exceptions to confidentiality provisions may be granted to those responsible for making decisions during arbitration, litigation, or other administrative functions arising from a positive drug test. Statistical data, records, and/or reports may be maintained by the employer/company and drug-testing laboratories. The statistical data are gathered to monitor compliance with rules and quality assurance/initiatives and to evaluate the effectiveness of the drug-testing programs or laboratories.

Preplacement Physical Examinations for New Hires

A history and musculoskeletal examination and a routine physical examination will be performed on all regular new hires to evaluate the ability to perform their job duties with respect to employer's/company's job demand needs and the ADA. A personal history and self-assessment form will be completed (e.g., latex sensitivity questionnaire for new-hire nurses). A drug-testing urine drug screen and other physical examinations as required will be performed on site (e.g., coke oven, benzene).

The typical examination, as stated in Chapter 1, includes the following:

1. Examination for musculoskeletal impairment and routine screening for other medical conditions such as high blood pressure or chronic diseases.
2. Routine screening investigations as indicated based on job title, job demands, and history of work. Procedures and services are limited (e.g., blood count, dipstick urinalysis, audiometry, chest x-ray, visual acuity, spirometry, and clinical chemistry). Special investigations or studies may be recommended.
3. Alcohol and drug testing for commercial drivers as it relates to DOT and the FHA.
4. Recommendations for workers who will be performing heavy lifting (40–60 pounds) or other stressful or unusual physical activities.

Diagnostic Reporting or Coding

Results of evaluations for both new hires and others (e.g., fitness evaluations) will be recorded in a format consistent with the employer's/company's data needs. An automated information system of reporting or coding the Department of Occupational Health's activities can easily be applied to the employer/company practice. Results of evaluation will be immediately available to the employee and supervisor in the activity prescription report, which is completed at the time of a clinic encounter. This form can be customized as well to meet the company's data needs (see Activity Prescription form—Chapter 1, Figure 1).

MEDICAL MONITORING AND TESTING REQUIRED BY LAW OR EMPLOYER/COMPANY POLICY AND/OR OCCUPATIONAL EXPOSURES

Medical Monitoring

Most on-site teams currently provide specialty evaluation and treatment, safety programs, prevention services, and regulatory compliance assistance to employers. Examples of the services available are listed in Table 2.

Occupational Exposures

Routine review of the Material Safety Data Sheet (MSDS) is necessary to identify possible exposures (e.g., new paint). The OSHA 101 and 200 logs will be maintained and reviewed. Job titles and descriptions will be reviewed in terms of physical demands and environmental conditions (fumes, odors, dust, gases, machines, equipment, etc.) that may pose risk to employees. The on-site physician will examine employees who have experienced a hazardous exposure (e.g., asbestos) and will perform appropriate screening evaluations. The on-site model in-

TABLE 2 Clinical Activities for Safety and Regulatory Compliance Administrative Activities

Clinical activities	Administrative activities
Respirator fit	Eye protection education
Noise-exposure evaluation	Chemical spills education
Hearing conservation	Cumulative trauma education
Asbestos-exposure monitoring	Hazard communication
Bloodborne pathogen testing and treatment	Record keeping (Material Safety Data Sheet (MSDS), OSHA 101 and 200
Lead testing and exposure evaluation	

Source: Courtesy of Gregory Preston, M.D.

cludes providing educational materials and assistance in writing policy and procedures pertaining to safety and compliance at the company.

Tests Required by Regulation

Evaluation and assessment programs will be conducted at regular intervals. Audiometric testing will be conducted annually, and a periodic medical evaluation of the respiratory protection program will be performed on site. The on-site team will track the progress of required surveillance testing and will notify plant management, safety, and human resources of the status of the programs. If the plant cannot complete the schedule of required testing, the on-site team will work with the company to obtain additional services from an authorized provider or member of the on-site team (e.g., occupational health nurse or audiologist).

MEDICAL EVALUATION OF EMPLOYEES WITH WORKERS' COMPENSATION CLAIMS

The occupational medicine case manager (see Appendix 1) works with workers' compensation representatives, job-placement coordinators (both union representatives and management representatives), and regional personnel administration or human resources to evaluate placement, total and permanent disability status, average total claim costs, paid indemnity costs, paid medical costs, unmanaged claims, open cases, and transitional work guidelines. Despite initial union resistance, the case manager is able to get employees back to work. Anticipated time off of work due to surgery and light-duty assignments are managed. Outlier management, employer involvement, and worker input is the evolution of workers' compensation managed care. Later, the sickness and accident (S&A) employee claims can be brought into this process. As stated earlier, it is important that productivity be maintained. Continuing care will be monitored or coordinated with the worker's family or personal physician. ''Symptom magnifiers,'' or patients/workers who are less than motivated to return back to work, are managed through objective findings and case review.

The occupational medicine physician will oversee and medically manage all work-related injuries and illnesses. Specialty care services require a referral from the occupational medicine physician team, which includes company case managers and agents, and communication with the claimant's attending physicians.

It is recommended that the on-site health provider team and the employer offer a broad range of board-certified specialists for independent medical observations (IMOs) and independent medical evaluations (IMEs). Immediate access to physicians should enhance the company's mission of giving prompt attention to returning its employees to duty as quickly as medically indicated. Other diagnostic and treatment services can be linked to the computer for scheduling and for obtaining medical record reports.

Often tests require days to complete, but that is not the case with effective case management. The occupational health team recognizes that from the initial point of injury/illness the patient will need prompt diagnosis and treatment to prevent lost work time and to relieve patient symptoms. The major advantage to the on-site company or plant medical director and the department physicians is the availability of services that will improve the ability to make prompt decisions concerning the course of treatment.

The occupational medicine physicians will be available to company patients during the normal work schedule of the employee. There will be easy access to the physician and his or her team during all three shifts of operations, Monday through Sunday. The on-site physician will also carry a state-to-state pager at all times. Coverage will continue during vacations by the occupational medicine physician or replacement physician.

Other Services, Immunizations

Companies generally provide only tetanus immunizations. Allergy injections and travel immunizations are not customarily provided. At the company's request, the on-site team of providers will follow the recommendations of U.S. Department of Health and Human Service Centers for Disease Control as it relates to bloodborne pathogens standards administration, compliance, and consultation. All staff will receive annual education and training in bloodborne pathogens and vaccinations (e.g., hepatitis B immunization) offered to any high-risk group of workers. Other routine vaccinations such as influenza, pneumococcal, or hepatitis may be administered as selected by the corporate client. Executive physicals may require special arrangements for travel medicine.

Epidemiologic Surveillance

Incidence of infection (observation regarding the incidence of infections) due to tuberculosis or clusters or patterns of illnesses or risks shall be communicated to the plant manager and human resources. The on-site occupational health provider may be asked to participate in projects that develop and test best practices and partnership models (i.e., Food and Drug Administration research study to test cold laser treatment of carpal tunnel syndrome). Providers may agree to participate in a special health study or health trend analysis that may be scientific and arranged by the plant or human resources and conducted by consultant services to measure quality, cost, and outcome. Knowledge of an employee death due to an occupational illness or injury should be immediately communicated to the plant manager, and the appropriate plant policies and procedures should be adhered to. Proper medical directives should be known, and ongoing training and communication is required.

Health, Education, Health Promotion, Screening Programs

The on-site occupational health physician, nurse, physical therapist, occupational therapists may provide a professional opinion on a case-by-case basis for referral to employee assistance programs where they are in place in the plant. Preventive case management offered year round by the plant on-site physician to the individual who would like to make some lifestyle changes (i.e., weight loss, smoking cessation) but doesn't have time to attend a class or seminar is not uncommon. The focus is on those individuals with risk factors for cardiovascular disease. Individual one-on-one counseling options offered throughout the year can include:

First aid
Workplace stretching and strengthening exercises
Stress management
Reducing cholesterol
"Keep your back healthy"
Heart smart/nutrition
Weight Reduction
Arthritis/Joint protection
Control of high blood pressure
Correct lifting, bending, body mechanics, and tool use
Ergonomics and safety
Diabetes education

It is important to train, orient, and retain all personnel contracted to the company. The personnel shall receive all training that pertains to the company's contractual issues throughout the term of the agreement or direct company contract.

The relationship created by the on-site team with the company is that of an independent contractor, thus it is the responsibility of the on-site team to train and provide qualified health care replacement workers to cover absences, vacations, etc. Responsibilities include specific training in industrial, occupational, rehabilitative, and preventive medicine as well as audiometric, pulmonary, and regulatory compliance issues in order to obtain and educate the company on the most current information in the field.

Physician services costs are based on an hourly rate for the quoted straight time price. Overtime hours will not increase the quoted rate. Required medical staffing personnel will be provided by the on-site health care provider. The on-site team will be paid at the rates as set and agreed upon, with payment to be made monthly. The price is based on salary, benefits, and general practice insurance schedules as shown in Table 3. Staffing will be provided 7 days a week, days and afternoons are specified by the company. Customers requiring physician coverage during all three shifts, Monday through Sunday, will be quoted based on the number of physicians contracted for.

TABLE 3 Sample Costs for Occupational Medicine Physicians On Site

Number of staff and title	Price/Cost per hour
1 staff occupational medicine physician (1 FTE)[b]	1 FTE × $175,000 base salary[a] plus benefits at $35,000 and malpractice insurance at $12,250 *Total annual = $222,250*
1 staff physician (1 FTE), family practice, internal medicine, or emergency medicine	1 FTE × $145,000 base salary plus benefits at $29,000 and malpractice insurance at $10,150 *Total annual = $184,150*
1 staff physician (1 FTE), family practice, internal medicine, or emergency medicine	1 FTE × $145,000 base salary plus benefits at $29,000 and malpractice insurance at $10,150 *Total annual = $184,150*
1 staff physician (1 FTE)	1 FTE × $155,000 base salary plus benefits at $31,000 and board eligible in occupational medicine malpractice insurance at $10,850 *Total annual = $196,850*
	Total annual cost = $787,400
	Monthly cost = $65,167
	Monthly management fee = $9,843

[a] The staffing configuration represents cost by physician specialty on site.
[b] FTE = full-time equivalent.

For companies and employees, the purchasing or the buying of health care services is bringing the purchaser/supplier relationship together. The talent and creativity of the on-site team and the producers is adding value. The role of the physician and the ability to deliver the needed continuous improvement is understood by the employer. This concept is written into the physician job duties. See Appendix 2 for a sample job description of a staff occupational medicine physician.

The on-site team physician should provide services during normal business hours as stated by the company; however, extended hours may be available. Hours will be flexible to accommodate production shifts as required upon notice (preferably 72 hours).

The industrial rehabilitation team (i.e., physical therapists, occupational therapists, athletic trainers) should agree to provide appropriate staffing on all shifts when the local contract agreement is ratified to require such. The on-site industrial rehabilitation program includes the services of licensed and/or registered physical therapists and registered and board-certified occupational therapists, who in addition to specific injury treatment will provide workplace analysis, ergonomics, education and training of employees in the prevention of illness and injury, and a smooth transition from the formal rehabilitation program to a fitness/ wellness center and behavioral modification of lifestyle.

The advantages to the company include:

1. On-site industrial rehabilitation meets the Department of Public Health and the Joint Commission on Accreditation of Healthcare Organizations (JCAHO) accreditation standards and may be a part of a leading medical, education, and research facility.
2. Providing rehabilitation on site gives therapists a chance to monitor an employee/patient condition directly by evaluating and observing the employees' workstation. The therapist can structure therapy around the patient's actual work environment, creating an appropriate conditioning program to help minimize lost work time.
3. Postdischarge from the rehabilitation program, the therapist will instruct the employee/patient on the correct exercises to perform in the fitness center. The wellness process will encourage a lifetime of proper exercise with long-term follow-up to prevent injury.

Employers are the major purchasers of health care services. Their strong desire to reduce escalating health care costs, improve time off work, and improve the quality of the work life of their employees has led to requests for on-site rehabilitative services offering employers, workers' compensation/third-party administrators, self-insured companies, and ergonomic departments' assistance in meeting these needs.

The key quality characteristics in order to manage patient, employers, and insurance customer needs include timely evaluation, treatment, case management functions, and prevention.

Twenty-four (24) product attributes for on-site industrial rehabilitation are outlined below:

1. On-site rehabilitation consists of a registered, board-certified occupational therapist to perform on-site job analysis and to provide early detection and corrective treatment for cumulative trauma disorder (CTDs) in the workplace and to report recommendations or redesign of work stations and modifications of equipment. Splinting and hand therapy treatment is provided.
2. A registered physical therapist provides traditional physical therapeutic intervention with emphasis on aggressive treatment application to facilitate early return to normal activities of daily living. Lumbar MedX treatment, Biodex isokinetic, isometric, eccentric treatment, manual therapy, myofascial release, trigger point release, diagnostic

treatment and testing can be performed. Modalities utilized can include heat, cold, paraffin, ultrasound, whirlpool, iontophoresis, phonophoresis, and electrical stimulation, among others.

3. Employees/patients attend physical and occupational therapy prior to or after their work shift. Employees off work may attend during the regular hours.
4. On-site industrial rehabilitation consists of rehabilitative services offered to companies and their employees at work on the plant premises.
5. Contract, agreement, or preferred provider relationship with a local, national, or multinational corporation.
6. Reduction of cumulative trauma disorders through the implementation of control strategies to reduce and eliminate OSHA-cited trauma disorders in the workplace.
7. Formation of an interaction between on-site industrial rehabilitation staff and the company's regional personnel administration; union, medical department; workers' compensation representatives; benefits representatives; the National Joint Health and Safety Committee and company representatives; ergonomics representatives; workers' compensation claim processors; and plant or company management; case managers and job-placement coordinators.
8. Increase in the number of employees who are injured or have illnesses that *remain working*, thereby reducing time off. This facilitates productivity in the earlier phase of their recovery.
9. Reduction of litigation cases and elimination of employee medical restrictions.
10. Reduction of absenteeism because employees are not isolated from workers.
11. Reduction in workers' compensation claims, along with expenditures for replacing injured employees.
12. Increase in personal/professional responsiveness on the part of the rehabilitation team is favored by the union through responsive communication and clinical evaluation.
13. Having licensed and/or registered physical therapists and registered, board-certified occupational therapists available on site to meet current and future needs of companies provides the highest-quality services at the least cost.
14. Prevention programs and proper reconditioning are provided. Programs are designed to reduce health care costs for employers and their employees without reducing quality of care.
15. "Grand rounds" or utilization review of workers' compensation cases are held monthly with all representatives of the medical management process present. Patient status in occupational medicine and

industrial physical therapy and occupational therapy are discussed. The clinical diagnosis and outcomes are reviewed to foster discussion, education, and rehabilitation goal attainment.

16. A joint medical advisory board should be set up, meeting quarterly. The medical advisory board is established to meet the Joint Commission on Accreditation of Healthcare Organizations and Medicare standards of medical direction for all patients treated on site. The primary purpose of the medical advisory board is to approve all policies and procedures having an impact on the care of patients, monitor patient outcomes, and assist the company in accomplishing goals that will enhance the delivery of care.

    Rehabilitation staff responsibilities include participating in development and measurement of quality, cost, and outcome. The revision of guidelines, quality assurance, infection control, and patient care is ongoing. This insures that the programs will be evaluated for appropriateness and effectiveness. Processes and process control include:

    A. Establishing criteria for the initiation and continuation of occupational health and rehabilitation programs.
    B. Evaluating existing patient care programs on a periodic basis in terms of quality, effectiveness, patient and medical staff satisfaction.
    C. Evaluating all proposed patient care programs in the same manner.

    Each staff member will function as a resource within the corporation and community to promote the scope of activity and quality provided to on-site patients (see Appendix 3).

17. Workers' compensation representatives approve the prescriptions for therapy treatment and personally sign all orders for therapy prescriptions.
18. "Passive" surveillance of employees/patients will be available, and predictions will be made about employees'/patients' ability to return to work.
19. Cost-saving reports, employee work location, outcome studies, insurance mix analysis, and quality assurance reports are kept. The data are reported to the company to monitor health care savings and to compare performance with a broad cross section of specialty providers.
20. Employees/patients are kept on site to avoid therapy treatment patterns or levels that are excessive or inappropriate. The company physician is on site for consultation and communication with the therapist.

21. The continued success of the employer's/company's managed health care program depends on its ability to maintain and build a panel of high-quality medical staff.
22. Travel distance between work and the rehabilitation center is decreased and waiting lists for appointments are shortened.
23. Rehabilitation is subject to the collective bargaining agreement or employee contract.
24. Rehabilitation services are provided in compliance with application laws, including statutes and regulations related to operations.

PROOF OF MALPRACTICE INSURANCE

Physical and occupational therapists malpractice insurance (professional liability insurance) is provided.

The rehabilitation staff shall have insurance (professional liability), on such terms and in such amounts as shall be satisfactory to cover medical staff for all duties performed within the scope of the agreement with the employer/company (i.e., the automotive assembly plant). The rehabilitation provider will give the employer/company evidence of the malpractice insurance (professional liability insurance) policy.

Prior to commencement of services hereunder, and during the term of an agreement and any extensions or renewals thereof, the rehabilitation staff provider shall:

(a) Obtain and maintain at its sole expense insurance coverage or self insurance as set forth in the example listed below:

Comprehensive general liability including blanket contractual and fire legal liability coverage	$5,000,000 combined single limit for personnel injury and property damage
Automobile liability covering all owned, nonowned, and hired vehicles	$2,000,000 combined single limit for personal injury and property damage
Medical malpractice	$10,000,000 per occurrence/$19,000,000 aggregate
Workers' compensation	Statutory
Employers' liability	$250,000

(b) The on-site rehabilitation provider shall furnish to the employer/company certificates of insurance executed by its insurer or insurers evidencing that the above insurance is in full force and effect.

(c) The on-site rehabilitation provider's purchase of insurance coverage or the furnishing of certificates shall not release the rehabilitation provider of its obligations or liabilities under the agreement. The on-site provider hereby grants to the employer/company access to all books, receipts, records, written instructions, and correspondence related to work authorized under this agreement as required by law.

Benefits and savings to employers are as follows:

1. There will be a decrease in employee absenteeism and a decrease in the cost of workers' compensation expense.
2. On-site job analysis will reduce accident frequency.
3. Productivity will remain high.
4. The company will be responding to OSHA regulations.
5. Case management and follow-up on lost time injury/illness will be provided.
6. Personal/Professional responsiveness on the part of clinical staff will be favored by the union.
7. Savings will be calculated as the difference between the amount billed and the amount for payment (i.e., state maximum hospital payment ratios set by the Bureau of Workers' Compensation). Also, fee discounts will be compared to market prices.

REFERRAL PROCESS

1. Contact with the employee/patient will be within 24–48 hours of the referral for rehabilitation services.
2. Clear, accurate, and timely communication will be mandatory. Written reports and recommendations will be forwarded to the company within 24 hours of the initial evaluation by rehabilitation staff.
3. There will be close coordination of services with rehabilitation personnel.
4. A job analysis will be performed as needed.
5. Grand rounds or utilization review of workers' compensation cases will be held monthly.
6. Statements/Billing of services rendered with accompanying reports will be sent monthly.

For the company and the health care provider, the ability to form a structure to act on and understand the tough issues of managing workers' compensation care is critical. The dialogue between health care provider and the leadership increases the continuous improvement cycle, and the link between the employer and the delivery team is made stronger.

APPENDIX 1: OCCUPATIONAL MEDICINE CASE MANAGEMENT FUNCTIONS

Occupational medicine case management consists of actual direction and management of the injured or ill employee's/patient's care. A minimum of 5 years of experience and specialized training in this field is required. Other requirements are:

1. Clinical experience in dealing with work related illnesses and injuries.
2. Knowledge of OSHA, state, Bureau of Workers' Compensation, and company specific reporting requirements.
3. Ability to identify early appropriate care; knowledge of who can deliver prompt initial treatment (urgent, emergency department, follow-up, rehabilitation, and surgery) within the medical centers, hospitals, and the community.
4. Ability to identify testing requirements, such as diagnostic services (i.e., electromyography), active rehabilitation, time-limited and outcome-oriented physician services; ability to monitor effective utilization of these procedures.
5. Ability to communicate and coordinate medical reports within specific time frames, including working with company workers' compensation claims staff and billing departments to process claims rapidly and accurately; ability to communicate with employees, employers, insurance company, workers' compensation representatives, benefits representatives etc.
6. Knowledge of union laws and collective bargaining pertaining to ADA, restricted work, on-site job analysis, ergonomics, and workplace safety standards.
7. Ability to identify employee/patient capability and work with company functional job descriptions to return employees/patients to work or to light duty jobs (this avoids litigation and reinjury); ability to work with medical/rehabilitation team.
8. Cooperate with company disability management programs to resolve employee/patient lost work time; discuss specific return to work guidelines and modified programs for injured workers.

9. Focus on injury prevention, quality care, wellness, and education programs to minimize reoccurrence.
10. Customize and implement payment arrangements with employers such as managed care programs to hold costs down and to encourage efficient care and superior return to work results. Collaborate with other departments (i.e., Labor Affairs, the Office of Business and Labor Relations, and the Managed Care Department) on company-specific opportunities.
11. Familiarity with occupational health practices (i.e., self-insured employers, third-party administrators, and worker tax-free wage-loss benefits that can meet and sometimes exceed regular wages).
12. Ability to develop guidelines for outcome research and reporting, refine and revise, and be accountable.

APPENDIX 2: POSITION DESCRIPTION FOR OCCUPATIONAL MEDICINE PHYSICIAN

Position Summary

The occupational medicine physician (OMP) is responsible for the medical direction of the occupational health department through the management of the delivery of occupational and environmental medicine provided to employers under contract to their employees.

Key Tasks and Responsibilities

The OMP must be able to:

1. Insure the services provided are consistent with good clinical care and in keeping with contract obligations.
2. Work on-site at company locations performing occupational health services, which include preplacement/new hire exams, employability determination, transitional work opportunities, sickness and accident medical leaves, annual infection control, bloodborne pathogen program, and overseas immunizations.
3. Detect and treat workers' compensation and sickness and accident reported injury and illness in the workplace.
4. Coordinates care, including support and assistance in regulatory compliance [National Institute of Drug Administration (NIDA) drug screen, OSHA, ADA, Medical Review Officer (MRO)] and streamline processes for treating clinicians.
5. Participate in a clinical consulting network, which will provide high-quality care and fast response to program needs (grand rounds, company workers' compensation and benefits representatives, ergonomics team).
6. Promote continuous quality improvement using data related to outcome assessments, practice guidelines, and other methods.
7. Apply knowledge of toxicology, epidemiology, ergonomics, biostatistics, and related disciplines of occupational and environmental medicine. (The performance goals of the OMP are shown in Table 4.)

Table 4 Performance Goals for Occupational Medicine Specialist (Staff Physician)

Goals	Moderate progress	Fully achieved
Clinical		
Participate as a clinician to provide high-quality care and fast response to occupational health program needs	Provide clinical coverage in direct patient contact	Review all workers' compensation cases currently active, evaluate patients, make treatment recommendations, and identify individuals who no longer are completely disabled and may return to work to limited duty or without restrictions
Detect and treat workers' compensation claims for work-related sickness/accident and illness/injury, and disability management	Provide clinical coverage to one or more Department of Occupational Health employer contract sites	Provide independent medical coordination and support to a Department of Occupational Health contract site
Administrative		
Provide clinical direction to other (nonphysician) providers at a department of occupational health contract sites	Apply knowledge of specialty to achieve regulatory compliance for activities conducted at a department of occupational health contract sites	Streamline clinical processes (e.g., needle stick, respiratory fitting) for clinical providers working at Department of Occupational Health contract sites
Promote continuous quality in occupational health delivery by the application of knowledge of toxicology, epidemiology, ergonomics, biostatistics, and related disciplines of occupational and environmental medicine	Provide orientation and training to primary care providers assigned to occupational medicine contracts	Develop two new programs in occupational health (e.g., practice guidelines, health and safety, and ergonomics)

8. With support of information services, establish an information-management system that will facilitate the prompt and efficient management of cases, their tracking, and eventually a database that will allow determination of reasonable expected outcomes for a variety of clinical conditions. The data collected will cover not only clinical outcomes, but associated costs as well.

Contact with Others

The OMP works closely and communicates frequently with the administrative director of the department of occupational health and with other physicians and nonphysician clinical and management staff of the program. The OMP communicates frequently with executives and line managers of companies that contract for services and with health care providers to whom employees of the contract company might be referred.

Reporting Relationship

The OMP reports to the administrative director of occupational health for administrative matters and the plant manager or on-site customer.

Qualifications

The following qualifications are required:

Board certification as a specialist in preventive medicine, completion of a residency program in occupational and environmental medicine. Three years' experience in an occupational health setting.

Strong interpersonal skills and ability to communicate effectively, both orally and in writing, with persons from a variety of backgrounds.

A commitment to continuous quality improvement and team-oriented leadership.

APPENDIX 3: DEPARTMENT OF OCCUPATIONAL HEALTH INDUSTRIAL REHABILITATION PROGRAM

Medical Advisory Board

The on-site medical advisory board was established to meet the Joint Commission on Accreditation of Health Care Organizations and Medicare standards of medical direction for all patients treated within the occupational health and industrial rehabilitation system. The primary purpose of the medical advisory board is to approve all policies and procedures having impact on the care of patients, monitor patient outcomes, and assist the on-site health care provider to accomplish goals that will enhance the delivery of care to our patients.

These responsibilities include participating in the development, measurement of outcomes, and revision of such committees as quality assurance, infection control, and patient care. This insures that the board will evaluate the appropriateness and effectiveness of the occupational health and industrial rehabilitation patient care programs by:

A. Establishing criteria for the initiation and continuation of rehabilitation programs
B. Evaluating existing patient care programs on a periodic basis in terms of quality, effectiveness, and patient and medical staff satisfaction
C. Evaluating all proposed patient care programs in the same manner

Each board member will function as a resource within our corporation and community to promote the scope of activity and quality provided to our patients.

Occupational Health and Industrial Rehabilitation Services Mission Statement

The occupational health and industrial rehabilitation mission is to be positioned as (1) providing high-quality care, (2) delivering appropriate, comprehensive services to people who need rehabilitation, (3) cost and time efficient, (4) highly effective in terms of outcomes, and (5) convenient for patients.

To be successful, occupational health and industrial rehabilitation recognizes the importance of remaining on the forefront of medical technology and health care delivery, preventing disease and promoting wellness, and being a responsible corporate citizen.

Philosophy

The on-site occupational health and industrial rehabilitation medical and administrative staff are dedicated to the provision of quality care. Interest in documenting this quality has prompted the team to incorporate existing functions in the quality management program directed by the occupational health and industrial rehabilitation medical advisory board. The program shall allow for the participation of all medical staff departments, corporate committees, subcommittees, ad hoc committees, and activities as identified.

Responsibilities

The occupational health and industrial rehabilitation quality management program necessitates the active participation of the company, the medical staff, administration, and other professional staff as outlined in the Joint Commission on Accreditation of Healthcare organizations and other licensing and accrediting standards.

The Company

The company has the ultimate responsibility for the quality of care rendered. The company has delegated specific functions to the medical staff and administration to be conducted under the direction of the on-site occupational health and industrial rehabilitation medical advisory board committee and coordinated through the on-site provider. The company reserves to itself and to its patient care committee the authority to:

Receive and review, accept or reject periodic reports of the program activities
Act on its own initiative when necessary

The occupational health and industrial rehabilitation advisory board shall assist the company in the interaction and evaluation of the quality-management activities of the program.

The Medical Staff

The medical staff is responsible for implementing the program as it concerns medical services and patient outcomes. Each chairperson or medical director shall be responsible for the quality of patient care rendered by occupational health and industrial rehabilitation services and shall participate in the program in accordance with the Joint Commission on Accreditation of Healthcare organizations and other licensing and accrediting standards. Each medical staff committee member shall carry out its functions in a manner which shall enhance the success of the occupational health and industrial rehabilitation program.

Administration

The company plant manager or president and the on-site health care provider, through the existing organizational structure, are responsible for implementing the program as it concerns administrative services. They shall:

Implement the quality management program as it concerns nonphysician professionals and technical staff
Provide necessary administrative assistance to support and facilitate the continuing operation of the program
Analyze information and act upon problems involving technical, administrative, and support services and hospital policy
Establish reporting mechanisms so that findings and recommendations from quality-management activities are shared between medical and administrative staffs.

Professional Practice Executive Committee

The company occupational health and industrial rehabilitation medical advisory board is comprised of members of the medical and administrative staff of the company and on-site health care team and shall:

Receive and review, accept or reject, the findings, actions, and outcomes of quality-management activities from its sectional committee
Make additional recommendations or act to resolve identified problems not previously or appropriately addressed
Determine the direction for dissemination of informing to appropriate individuals or departments for institutional problem resolution
Prioritize institutional issues
Conduct annual program evaluation

Meetings

The company occupational health and industrial rehabilitation medical advisory board shall meet quarterly (or as frequently as necessary) to assure efficient and effective program operation. A quorum, defined as 50% plus one of the committee's voting membership, will be required to conduct business.

The On-Site Quality Assurance Program

The on-site quality assurance program is an administrative unit and shall serve as the focal point for the coordination and documentation of the program. The administrator shall:

Inform the company medical advisory board regarding quality management requirements, regulations, and standards
Recommend mechanisms for quality management development, implementation, improvement, and problem resolution
Review, evaluate, and display data relative to the quality-management program
Coordinate and integrate the collection and flow of data for the program
Track and monitor identified problems and action taken
Act as liaison to external agencies requesting information on quality-management activities or providing data to the hospital

Methods

The medical advisory board will use a variety of ongoing methods to assess the quality of care and professional practices of the occupational health and industrial rehabilitation program. The methods shall include, but not be limited to, the review of patient care records and documents, patient satisfaction surveys, observations, interviews, complaints, reports of unusual occurrences, and department reviews.

Reporting

An ongoing reporting system shall be maintained to assure that the appropriate individuals or groups are informed of those quality-management activities appropriate to their functions. Reporting to the quality assurance administrator will be accomplished through a variety of methods including, but not limited to, minutes, quarterly reports, infection-control surveillance documents, and equipment safety check reports.

All communications and reports documenting the integration and interaction of those resources stated above shall be maintained by the on-site occupational health and industrial rehabilitation administrator.

The company's medical advisory board shall provide periodic reports (at least quarterly) of their findings and actions.

Confidentiality

The company and on-site health care team as delegated to the medical staff and administration through the professional medical advisory board shall review the professional practices and quality of care, in accordance with the provisions of the state public health code and any other statutory or regulatory authority requiring or providing for such reviews. Such reviews shall be for the purpose of reducing morbidity and mortality and improving the quality of patient care provided in the system.

Confidentiality is essential to the effective functioning of individuals, committees, or departments performing these functions. Candid and thorough evaluations of professional practices and patient care are essential to the improvement of professional practices and patient care. Consequently, all records, data, and knowledge collected for or by individuals, committees, or departments performing this professional review function are and must be confidential. Such records, data, and knowledge are not public records and are immune from subpoena as provided for by the state public health code and any other statutory or regulatory authority requiring or providing for such reviews.

The quality assurance administrator shall maintain all minutes, reports, worksheets, and other data in a manner ensuring strict confidentiality.

Minutes and Records

It is essential that quality-management activities, reports, recommendations, and data be kept confidential to ensure proper functioning of the medical advisory board duties. In order to maintain the privilege extended by law to such activities, certain safeguards must be maintained. Therefore, the minutes of the on-site medical advisory board and other reports and communications shall be filed by the occupational health and industrial rehabilitation administrator and filed by those committee individuals who have direct responsibilities relating to the activities of the program.

All files, reports, minutes, and other documentation shall be available to other appropriate internal or external agencies in a manner as to ensure that the privilege against disclosure or violation of privacy rights shall be maintained. Such material shall *not* be duplicated or distributed for inclusion in the files of these committees, departments, or external agencies.

Program Reappraisal

The structure and functions of the quality management program of the on-site occupational health and industrial rehabilitation medical advisory board shall be reappraised at least annually to assure the program is achieving its objectives, is effective, is cost-efficient, and is consistent with external requirements. This reappraisal process shall take into account the:

Evidence of impact
Resources expanded in carrying out program activities
Other internal and external information reflecting the program effectiveness

3

Disability Costs to Private and Public Programs

Physicians across the country are asked to extend accident and sick leaves longer than is medically needed. Specialty physicians are consulted, including orthopedic surgeons, neurologists, neurosurgeons, occupational medicine specialists, and plastic surgeons, among many others. Guidelines or ''activity prescriptions'' are utilized to assist the physician in writing job restrictions (e.g., no use of right hand) and prescribing the length of restrictions (e.g., 2 weeks). Patients are made aware of how long they will be allowed to rest the injured part or miss work for a particular malady.

Job placement coordinators are utilized in companies (e.g., an automobile plant) to help injured/ill employees return to work sooner in a transitional job that is less demanding prior to return to the original job and/or department. Job placement coordinators consist of both union and salary counterparts who are familiar with the jobs and duties available based on seniority. ''No job available'' becomes apparent when the patient/employee cannot perform all of the job duties or tasks of a particular job or an employee with greater seniority and no restriction has filled the job. Registered nurse case managers are part of the on-site team to retrieve physician evaluation reports and activity prescription guidelines. During contract negotiations, bargaining on behalf of union members includes provisions that employers address health and safety issues at the workplace.

On-site fitness, rehabilitation, and occupational medicine programs are being written into bargaining agreements. As the work force becomes older and more conscious of the benefits of prevention and ergonomics, on-site fitness centers and rehabilitation are increasingly utilized.

TABLE 1 Sample Cause and Associated Costs of Work-Related Accidents at a Financial Company

Cause	Number of claims	Dollar cost	Percent of total dollar cost
Stress (mental)	16	$256,302	23.8%
Overexertion (manual material handling)	73	$168,580	15.7%
Fall—same level	161	$167,744	15.6%
Bodily reaction	84	$142,917	13.3%
Struck by	174	$46,986	4.4%
All others	393	$293,613	27.25%
Totals	901	$1,076,142	100.0%

REPORTS: LOSS CONTROL AND COST SAVINGS

The on-site model can be sold to employers/companies identified by both the on-site provider and the employer. Table 1 and Figures 1–5 show the components of a loss control survey report. In reviewing the report, notice that the survey is based upon conditions observed and information supplied during the initial visit to the potential new customer company for proposing and implementing the on-site model in the service industry.

The information contained in this report is from a sample study of all employee injuries at a service industry corporation from January 1, 1997, to February

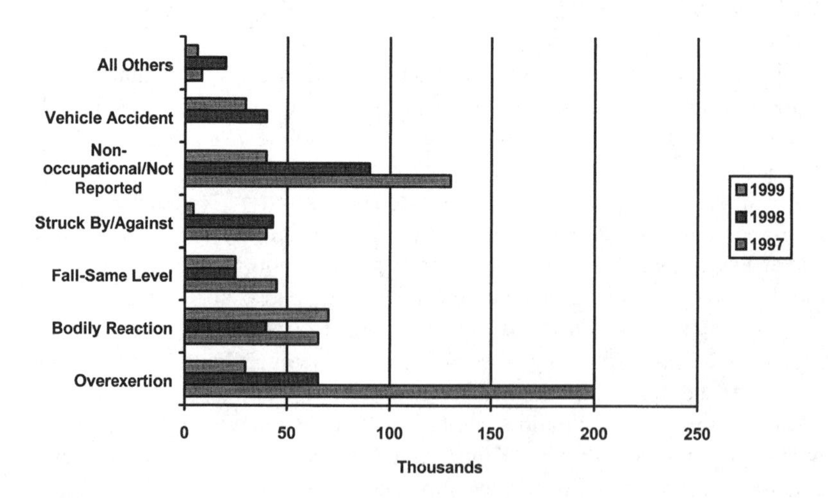

FIGURE 1 Claims cost by interaction, 1997–1999.

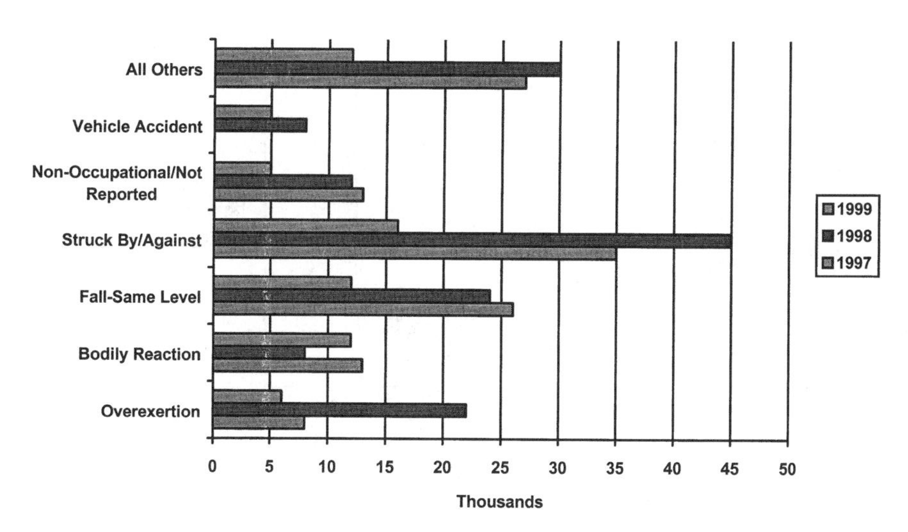

FIGURE 2 Claims frequency by interaction.

23, 1999. The dollar amounts listed are for "total incurred" (paid plus reserved) costs as of February 23, 1999. The focus of the study was the major causes of injury rather than the parts of body injured or the nature of the injuries. By correct identification of accident causes, positive steps can be taken to correct unsafe actions or conditions and to promote a safe work environment. The number of

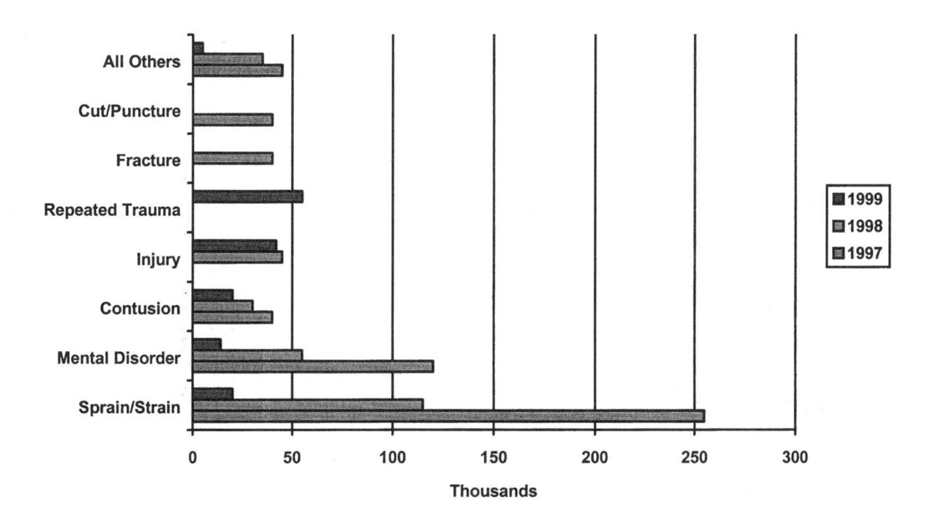

FIGURE 3 Claims cost by result, 1997–1999.

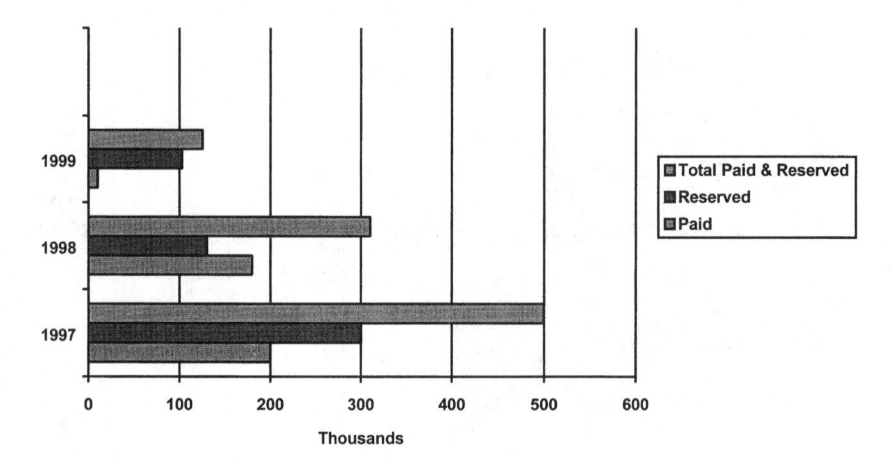

FIGURE 4 Total incurred costs, 1997–1999.

employees working at the company remained relatively (±20) constant. The 1999 claims could well exceed 1998 and 1997 based on the data presented. The service industry corporation was asking for a health care provider to decrease their exposure to costly claims. The on-site model was proposed.

As you can see by this analysis, the largest cause of dollar loss to the company can be attributed to injuries involving job stress. Stress is the direct

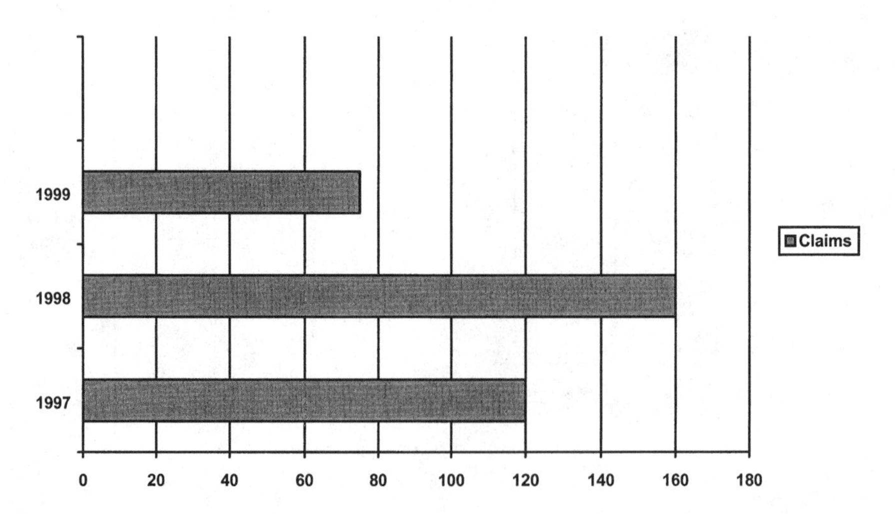

FIGURE 5 Claim rate per year, 1997–1999.

cause or very closely associated with 16 claims for a total of $256,302. The majority are complex claims currently in litigation.

The second highest cause of dollar loss is from injury suffered during manual material-handling activities (e.g., lifting, carrying, pushing, and pulling). No one employee group can be singled out as having a higher ratio of injuries. The on-site team addresses each of the major causes of loss and make recommendations to help reduce those losses in the future.

Stress-related claims at a company have proven to be low-frequency/high-severity type claims, with the average claim costing $16,000. It is very difficult to predict at what stress level employees become mentally overtaxed. This level varies greatly and is dependent upon many different inputs, large and small. Employees will experience stress as a result of basic job requirements and interaction with supervisors or managers. Stress sources off the job are numerous and sometimes interrelated with on-the-job stress. Most stress and psychological work claims are denied except in cases of violence or loss (e.g., bank robberies, airplane crashes).

Using data collected from the Bureau of Labor Statistics Economic Analysis in 1996 as well as historical data analysis in the six-state region—Illinois, Indiana, Michigan, Minnesota, Ohio, and Wisconsin—Table 2 shows the number of nonfatal occupational injuries and illnesses involving days away from work by part of body and selected natures of injury or illness in 1996. The awareness of the nature of the injury or illness (e.g., carpal tunnel syndrome) and the source of the injury (e.g., worker motions, floor, hand tools, chemical products, machinery, parts and materials, contact with objects such as struck by or against) will lead to safety and prevention programs to lessen the number of incidences.

Large payouts and settlements to employees who can no longer perform a job in a plant may be less costly for the employer than paying 80% of wages. Injured or ill workers give up the right to sue the employer in return for settlements. Payments of medical expenses and wage replacement is for life once the claim is accepted as compensable. The objective of the on-site medical team is to provide expedient and quality medical care and rehabilitation services. Workers' compensation insurance is mandatory in virtually all states. The on-site occupational medicine model works best with large or small (27,000 or 200 employees) companies who are self-insured. These employers, rather than purchase insurance from a commercial carrier, set aside a reserve to insure themselves against losses. The self-insured employer can either administer internally its own workers' compensation program or contract with a third party administrator (TPA) to manage workers' compensation claims for the company. Indemnity involves payment of wages for time lost, which is usually a percentage of the employee's gross weekly wage. The benefit is tax free. If the injured/ill employee returns to work part time or at a lesser hourly rate, the rate at the time of injury is applied.

TABLE 2 Occupational Injury Data

Part of Body	Part of Body Code 2	Total Cases	Nature of Injury or Illness 3				
			Sprains, strains	Fractures	Cuts, punctures	Bruises	Heat burns
Upper extremities	3	425,649	69,859	47,210	113,429	35,273	15,916
Upper extremities, unspecified	30	733	92	—	62	—	—
Arm(s)	31	80,153	21,698	10,266	8,966	11,649	3,865
Arm(s) unspecified	310	30,935	9,669	3,974	4,135	2,710	2,043
Upper Arm(s)	311	4,487	2,075	800	204	299	104
Elbow(s)	312	27,306	6,575	3,733	914	6,230	90
Forearm(s)	313	13,715	2,355	1,400	3,510	2,121	1,414
Multiple Arm(s) locations	318	2,797	760	254	131	200	73
Arm(s) n.e.c.	319	914	264	106	72	90	140
Wrist(s)	32	94,954	28,414	11,932	3,932	3,073	439
Hand(s), except fingers	33	75,610	6,656	5,615	25,667	10,607	7,227
Finger(s), fingernail(s)	34	152,585	8,612	18,823	73,326	8,456	1,596
Multiple upper extremities locations	38	21,271	4,262	556	1,468	1,437	2,784
Multiple upper extremities locations,		—	—	—	—	—	—
unspecified	380	171	—	—	—	—	—
Hand(s) and finger(s)	381	4,969	312	163	1,010	308	1,090
Hand(s) and wrist(s)	382	5,350	1,269	66	121	419	815
Hand(s) and arm(s)	383	3,444	366	84	100	307	568
Multiple upper extremities locations, n.e.c.	389	7,337	2,287	233	227	392	304
Upper extremities, n.e.c.	39	343	123	—	—	19	—
Lower extremities	4	375,538	161,950	50,357	22,356	66,105	5,626
Lower extremities, unspecified	40	306	134	72	—	—	—
Leg(s)	41	183,731	84,310	10,312	12,383	34,045	2,129
Leg(s), unspecified	410	28,515	7,979	3,739	4,077	6,265	690
Thigh(s)	411	7,890	2,456	480	1,792	1,473	844
Knee(s)	412	127,623	69,604	3,577	4,130	22,901	122
Lower Leg(s)	413	12,237	2,867	1,954	2,031	2,619	392
Multiple leg(s) locations	418	5,480	951	192	43	397	—
Leg(s), n.e.c.	419	1,986	453	370	309	389	68
Ankle(s)	42	86,713	62,687	13,047	630	4,163	424
Foot(feet), except toe(s)	43	68,355	10,458	15,195	7,958	19,182	2,568
Foot(feet), except toe(s), unspecified	430	61,823	9,856	13,488	6,825	17,645	2,332
Instep(s)	431	445	40	138	—	199	19
Sole	432	4,619	421	1,114	1,028	1,044	161
Sole(s), unspecified	4320	740	—	—	564	63	—
Arch(es)	4322	98	—	20	—	—	—
Heel(s)	4323	3,711	387	1,092	434	950	139
Sole(s), n.e.c.	4329	17	—	—	—	—	—
Multiple foot(feet) locations	438	240	49	71	—	63	—
Foot(feet), n.e.c.	439	1,228	93	384	86	232	52
Toe(s), toenail(s)	44	23,375	572	10,797	1,189	5,520	57
Multiple lower extremities locations	48	12,596	3,676	932	191	3,086	442
Multiple lower extremities locations, unspecified	480	164	50	—	—	—	18
Foot(feet) and leg(s)	481	1,276	144	145	—	332	171
Foot(feet) and ankle(s)	482	4,622	1,499	180	15	1,534	165
Foot(feet) and toe(s)	483	1,375	130	209	41	457	17
Multiple lower extremities locations, n.e.c.	489	5,160	1,853	383	123	731	72
Lower extremities, n.e.c.	49	462	114	—	—	84	—

Nature of Injury or Illness 3									
				Multiple traumatic injuries and disorders			Back pain and pain except		
Chemical burns	Amputations	Carpal tunnel syndrome	Tendonitis	Total	With fractures, burns, and other injuries	With Sprains and bruises	Total	Back pain, hurt back only	All other natures 4
1,656	9,807	29,924	12,727	10,019	3,791	1,634	18,315	—	61,518
—	—	—	114	16	—	—	—	—	386
567	42	—	4,129	1,329	270	312	5,495	—	12,145
366	17	—	1,126	397	85	102	2,138	—	4,360
—	—	—	155	62	15	—	179	—	595
—	—	—	2,045	516	81	128	2,302	—	4,893
141	15	—	540	188	—	—	419	—	1,610
46	—	—	218	148	56	48	394	—	572
—	—	—	45	18	—	—	61	—	114
52	—	29,924	4,994	543	193	230	4,518	—	7,132
860	114	—	1,414	1,591	409	163	3,654	—	12,205
114	9,631	—	561	4,545	1,989	296	2,231	—	24,690
64	20	—	1,513	1,972	915	624	2,347	—	4,848
—	—								
—	—	—	—	—	—	—	16	—	68
—	20	—	127	494	225	—	294	—	1,137
—	—	—	372	806	557	197	471	—	1,008
—	—	—	154	184	—	160	599	—	1,055
19	—	—	856	473	128	240	967	—	1,580
—	—	—	—	—	—	—	47	—	109
1,052	356	—	886	11,971	1,426	4,215	17,179	—	37,700
—	—	—	—	—	—	—	—	—	49
346	127	—	523	7,107	210	2,099	10,653	—	21,796
152	113	—	62	483	—	145	1,674	—	3,282
94	—	—	—	100	—	—	185	—	452
34	—	—	380	2,869	120	1,635	7,886	—	16,118
37	—	—	49	316	19	102	656	—	1,308
—	—	—	—	3,259	38	163	140	—	470
18	—	—	16	79	—	—	113	—	166
51	—	—	130	898	196	427	1,602	—	3,082
569	—	—	222	1,533	347	471	2,786	—	7,859
543	—	—	166	1,420	314	424	2,428	—	7,111
—	—	—	—	—	—	—	—	—	34
17	—	—	53	77	—	44	242	—	460
17	—	—	—	—	—	—	—	—	48
—	—	—	—	—	—	—	—	—	—
—	—	—	51	77	—	44	213	—	368
—	—	—	—	—	—	—	—	—	—
—	—	—	—	—	—	—	—	—	—
—	—	—	—	—	—	—	93	—	239
16	190	—	—	478	193	47	968	—	3,584
71	—	—	—	1,893	424	1,165	1,101	—	1,189
—	—	—	—	—	—	20	—	—	—
—	—	—	—	147	—	91	152	—	160
—	—	—	—	717	145	515	264	—	226
—	—	—	—	152	75	—	140	—	205
—	—	—	—	846	183	508	536	—	590
—	—	—	—	57	56	—	61	—	142

TABLE 2 Occupational Injury Data (*continued*)

			Nature of Injury or Illness 3				
Part of Body	Part of Body Code 2	Total Cases	Sprains, strains	Fractures	Cuts, punctures	Bruises	Heat burns
Neck, including throat	1	34,510	26,535	324	267	1,033	145
Neck, except internal location of diseases or disorders	10	34,510	26,535	324	267	1,033	145
Internal neck location, unspecified	11	52	—	—	—	—	—
Vocal cord(s)	12	48	—	—	—	—	—
Internal neck location, n.e.c.	19	40	—	—	—	—	—
Trunk	2	715,562	507,591	15,441	2,395	35,719	834
Trunk, unspecified	20	3,354	1,130	330	—	961	—
Shoulder, including clavicle, scapula	21	96,512	64,291	2,996	192	5,864	97
Chest, including ribs, internal organs	22	—	—	—	—	—	—
Chest, except internal location of diseases or disorders	220	29,031	9,086	5,747	591	8,602	386
Internal chest location, unspecified	221	158	—	—	—	—	—
Heart	223	568	—	—	—	—	—
Bronchus	224	1,096	—	—	—	—	—
Lung(s), pleura	225	2,131	—	—	—	—	—
Multiple internal chest locations	228	16	—	—	—	—	—
Back, including spine, spinal cord	23	490,608	398,063	3,235	728	11,351	142
Back, including spine, spinal cord, unspecified	230	221,510	181,254	1,293	176	5,143	74
Lumbar region	231	223,651	183,491	622	497	3,146	—
Thoracic region	232	18,290	15,162	120	47	420	—
Sacral region	233	4,397	3,618	60	—	165	—
Coccygeal region	234	4,006	276	872	—	2,072	—
Multiple back regions	238	17,298	13,381	248	—	366	—
Back, including spine, spinal cord, n.e.c.	239	1,456	882	19	—	38	—
Abdomen	24	41,173	6,439	71	354	1,018	78
Abdomen, except internal location of diseases or disorders	240	10,504	6,439	71	354	1,018	78
Internal abdominal location, unspecified	241	13,500	—	—	—	—	—
Stomach organ	242	1,064	—	—	—	—	—
Intestines, peritoneum	245	15,766	—	—	—	—	—

Lost wage payments (indemnity) continue for a lifetime. This is why often a settlement of a lump sum based on case review of the injured/ill employee's potential for maximum medical recovery is an acceptable practice. The case is closed once a settlement is accepted. Typically the indemnity payment exceeds medical costs. This is the area in which the on-site team can assist the employer in cutting costs. For example, listed below are the services that a 200-employee steel company selected for its on-site clinic:

Medical management of claims
On-site occupational health physician
  Physical therapist
  Occupational therapist

Nature of Injury or Illness 3									
				Multiple traumatic injuries and disorders			Back pain and pain except		
Chemical burns	Amputations	Carpal tunnel syndrome	Tendonitis	Total	With fractures, burns, and other injuries	With sprains and bruises	Total	Back pain, hurt back only	All other natures 4
—	—	—	16	228	—	189	2,830	—	3,107
—	—	—	—	228	—	189	2,830	—	2,932
—	—	—	—	—	—	—	—	—	52
—	—	—	—	—	—	—	—	—	44
—	—	—	—	—	—	—	—	—	40
656	—	—	2,984	4,909	342	3,500	69,268	52,046	75,763
—	—	—	—	64	—	—	377	—	429
—	—	—	2,900	908	—	773	7,969	—	11,289
—	—	—	—	427	137	194	2,423	—	5,703
66	—	—	—	427	137	194	2,423	—	1,700
—	—	—	—	—	—	—	—	—	156
—	—	—	—	—	—	—	—	—	568
—	—	—	—	—	—	—	—	—	1,096
—	—	—	—	—	—	—	—	—	2,131
—	—	—	—	—	—	—	—	—	16
44	—	—	—	2,243	—	1,709	52,046	52,046	22,754
—	—	—	—	979	—	695	22,993	22,993	9,571
17	—	—	—	956	—	790	24,772	24,772	10,139
—	—	—	—	18	—	—	1,744	1,744	752
—	—	—	—	—	—	—	338	338	193
—	—	—	—	38	—	—	480	480	263
—	—	—	—	229	—	163	1,392	1,392	1,655
—	—	—	—	—	—	—	328	328	182
454	—	—	—	38	—	—	1,393	—	31,324
454	—	—	—	38	—	—	1,393	—	659
—	—	—	—	—	—	—	—	—	13,500
—	—	—	—	—	—	—	—	—	1,064
—	—	—	—	—	—	—	—	—	15,761

Temporary modified work program
Safety and prevention classes
Ergonomic redesign of the work station

The company had seen a substantial increase in the number of time lost due to injury or illnesses. The president and chief executive officer had noted an increase in costs to the small plant. The author was able to demonstrate a decrease in time lost due to injury/illness percentage of the total workforce tracking results over 7 years. Each year the on-site model decreased the recordables (Table 3).

The on-site medical team working to case manage disability leaves of

TABLE 3 Michigan Occupation & Safety Health Administration Recordable Steel Company Frequency Report[a]

1992	1993	1994	1995	1996	1997	1998
533	417	389	122	35	23	7

[a] Lost time injuries/illnesses percentage of total workforce. On-site steel company contract was operational in 4th quarter of 1992. The number of employees remained constant at 200 until 1997–150 employees—and 1998–75 employees.
Source: Courtesy of J&L Specialty Steel Corporation.

claimants prioritizes cases or claims. The costs are decreased by the on-site physicians and therapists along with the grand rounds/case management team reviewing ''old'' or prior filed claims that have not been followed up for 6–18 weeks or longer and have no anticipated date of recovery or prognosis noted. A larger plant such as an automobile plant with a workforce of 3724 workers and disability leaves of 397 may demonstrate the types of disease or disorder by percentage of cases shown in Table 4.

TABLE 4 Automobile Plant Disease/Disorder Percentage of Workers' Compensation Claims

Diagnosis/Disorder	No. cases (%)
Nervous system in nature, including spinal cord injury, cerebral vascular accident, seizures, closed head injury	44 cases (11.1%)
Eye disorder	7 cases (1.8%)
Ear, nose, and throat	10 cases (2.3%)
Respiratory system (e.g., occupational bronchitis)	42 cases (10.6%)
Circulatory system (e.g., renal)	39 cases (9.8%)
Digestive system: hernia, gastritis, ulcers	24 cases (6.1%)
Hepatobiliary system and pancreas: cholycystectomy, alcohol/chemical/substance abuse	23 cases (5.8%)
Musculoskeletal system and nervous system and connective tissue: carpal tunnel syndrome; rotator cuff syndrome; knee conditions and procedures; epicondylitis; back pain, strain, and disc disorders; cervical neck pain; adhesive capsulitis; DeQuervain's tenosynovitis, fractures	175 cases (44.1%)
Skin, subcutaneous tissue, and cellulitis	12 cases (3.0%)
Hand trauma: amputations, wound, degloving, slivers or foreign objects, reflex sympathetic dystrophy, tendon lacerations	21 cases (5.3%)

Claims are reviewed for inappropriate restrictions (e.g., length of disability greater than 7–10 days and no available medical information). Repeat offenders or abusers with high absenteeism will be targeted. Patients with rotator cuff tears may be requested to perform a job that does not require "above-shoulder reaching." Communication with the injured/ill employee, management, and the union are key to the job placement process. The registered occupational therapist can evaluate job appropriateness and the ability of the patient/employee to return to work with restrictions.

The results or outcome measurements can be reported in dollars saved by not using outside medical services, reduction in lost work days or restricted work days, through transitional work placement, or the number of cases closed. Table 5 shows the savings in cost of medical services by providing an on-site program, as well as the savings reported in lost work days (i.e., placement in transitional work).

Employees need to be trusting of the on-site medical and rehabilitative team of health care practitioners. Physician and therapist board certifications, residency training programs, and affiliation with a major research teaching institute aids in the trust factor between patient and clinician. Any company or employer who does not have an on-site credentialed physician or therapist practicing has already set themselves up for high medical health care costs by not providing skilled professionals on site to manage and close the case. Outside referrals to specialists (e.g., orthopedics) can prolong the work restriction because the specialist is not familiar with the plant or job requirements and the associated physical demand characteristics. Computer input of data must be timely for analysis and follow-up.

The intent of the disability management program is to provide injured/ill patients/employees with meaningful work. The patient is invaluable to the company. The on-site model successfully returns patients/employees to work to establish a productive and profitable relationship between the company and its workers. Concerns for a quick and easy recovery are communicated to the employee.

TABLE 5 Cost Savings in Medical Services, Reduction in Lost Work Days Savings, and Savings in Return to Workforce or Transitional Work

1. *Dollar cost savings to plant (in medical services)*	
First year	$77,255
Second year	$119,587
2. *Reduction lost work days*	
First year	715 days
Second year	568 days
3. *Cases returned to workforce in transitional work*	
First year	44
Second year	150

For example, the United Auto Workers National Agreement for job placement of employees with disabilities has developed a process that is committed to enabling employees to be placed on jobs in line with their physical restrictions. Many patients/employees feel that they have recovered enough to work but not enough to perform their regular jobs. The occupational medicine physician on site will give the worker a well-defined restriction. The on-site team assists the worker in return to work. Seniority and other contractual considerations are adhered to in the job search process. The on-site physician and the patient's/employee's supervisor assist the patient in adhering to proper work restrictions until the condition improves. Disability leaves for nonoccupational sickness and accidents and occupational injury/illness are managed by the same processes and procedures.

The concept of tailoring cost savings reports is not unique to any one industry. The realities of workers' compensation costs is well known in the business community. Unfortunately, the only real way to successfully deliver care on site is to demonstrate the likely outcome.

Table 6 shows a sample cost savings report compiled for a service industry. The financial company contracted for on-site services at its two processing cen-

TABLE 6 Occupational Health Program: Workers' Compensation Loss Time/ Restricted Work Comparison[a]

Year	Lost work time (days)	Restricted work (days)
1998[b]	1042 (34% improv. from 1993)	1221 (54% improv. from 1993)
1997	963 (39% improv. from 1993)	1363 (49% improv. from 1993)
1996	592 (63% improv. from 1993)	530 (80% improv. from 1993)
1995	602 (62% improv. from 1993)	1177 (56% improv. from 1993)
1994	1428 (10% improv. from 1993)	1736 (35% improv. from 1993)
1993 Processing centers and 150 branches	1586	2663

[a] Savings in lost work days and restricted work days against 1993 when costs were escalating. The on-site model was contracted for and initiated in the 4th quarter of 1993. The number of full-time workers decreased by 350 in 1996 and increased by 100 in 1998 in the region reported.

[b] Ten severe cases accounted for 760 of the lost days or 73% of the 1998 total. 1998 numbers have increased due to complex injuries, including: knee fracture and surgery, shoulder rotator cuff tear and surgery, spider bite infection, severe back strain, and stress reaction/robbery.

ters. One center employed 1500 workers and the other 3000 workers. The 150 bank branch employees are also seen at the on-site occupational medicine and industrial rehabilitation clinic.

A bank branch typically employs 7 to 12 workers. Therefore, on-site space for medical and rehabilitative care is often not available. Injured or ill branch office workers are seen on site at the company processing centers. Payments or costs to private and public programs may be made according to a fee schedule set by the state or based on usual and customary charges. Protocols, disease management, or algorithm methods are used to control treatments (i.e., follow-up return office visits, the number of ancillary or diagnostic tests, and specialty services). The on-site model is applied to the branch office workers at the processing center sites and the same processes are adhered to.

The payment or fee arrangements can be agreed upon by the on-site provider and the company. The on-site provider bills according to the pricing mechanism selected by the company. Comprehensive managed workers' compensation program pricing and reimbursement strategies are as follows:

1. State Workers' Compensation Health Care Services Fee Schedule:
   A. Procedure codes were developed and copyrighted and are on file in the bureau of workers' disability compensation. The common procedure terminology (CPT) five-digit codes are utilized by providers to request for payment to the maximum fee. A written description including a bill or an attached document, which shall include a simple listing of services, the date of services, the procedure code, and payment, is requested.
   B. Facility service billing—all bills for facility services shall be submitted on the uniform billing UB-82 facility claim form. The state bureau of workers' disability compensation uniform billing manual shall be used in conjunction with the facility claim form. The state uniform billing manual assigns a maximum payment ratio to each facility based on a cost to charge.
2. Fee for Service (Nonoccupational): The on-site provider is reimbursed through the patients'/employees' insurance carrier. All operating costs are covered by the on-site provider. In the event a patients' insurance coverage does not cover all or part of the fees for a particular service, the patient will be responsible for paying the hospital when billed.
3. Company-Specific Contracts: This, the most widely used method in manufacturing, includes the process of bidding against other suppliers for on-site medical services. For example, worldwide purchasing departments request quotes for product specifications, including services offered, quantity, price, and delivery schedule. Purchase order terms and conditions are standard with all vendors. Collective bargaining

TABLE 7 Industrial Rehabilitation Center: Cost Savings in Travel Time, Mileage, and Treatment Modalities

	1990	1991	1992	1993	1994	1995	1996	1997	1998	Total
Total visits	2,794	4,121	4,708	5,255	4,688	5,032	3,912	3,587	2,218	36,316
Percent increase		47.5%	14.2%	11.6%	−10.8%	7.3%	(22.3%)	(8.3%)	(38.2%)	
Worker's Comp										
WC % of visits	55.00%	54.79%	69.34%	67.29%	61.55%	67.73%	49.90%	50.00%	46.32%	
WC visits	1,537	2,258	3,265	3,536	2,885	3,408	1,952	1,794	1,027	21,662
Percent change		46.9%	44.6%	8.3%	−18.4%	18.1%	(42.7%)	(8.1%)	(42.8%)	
Reduced travel time	$301,752	$445,068	$508,464	$609,580	$543,808	$583,712	$453,792	$416,092	$257,288	$4,119,556
Mileage savings—WC Patients	$15,370	$22,580	$32,650	$35,360	$28,850	$34,080	$19,520	$17,940	$10,270	$216,620
Reduced Modalities and Charges										
Outside facility	$343,327	$506,388	$578,519	$645,734	$576,062	$618,332	$480,707	$440,771	$272,548	
Industrial Rehabilitation	$79,908	$117,861	$134,649	$150,293	$134,077	$143,915	$111,883	$102,588	$63,435	
Net Reduction	$263,419	$388,527	$443,870	$594,441	$441,984	$474,417	$368,824	$338,183	$209,113	$3,423,778
Total cost savings	$580,541	$856,175	$984,984	$1,140,381	$1,014,642	$1,092,209	$842,136	$772,215	$476,671	$7,759,954

Assumptions

Travel Time:

4 hours per visit off site

Average salary and benefits = $27 per hour for 1990–1992

$29 per hour for 1993–1998

Calculation: Number visits (×) 4 hours/visit (×) average salary and benefits (per above)

Workers' Compensation mileage:

Visits = workers' compensation visits to industrial rehabilitation

Mileage = average 40-mile round trip (industrial rehabilitation to outside facilities)

Calculation: Number of workers' compensation visits (×) 40 miles (×) $0.25

Modalities and Charges:

Outside facility average charge per modality—$32

Outside facility average modalities per visit = 3.84

Industrial rehabilitation net charge per modality = $13

Industrial rehabilitation average modalities per visit =2.2

Calculation: Total visits (×) $32 (×) 3.84 modalities for outside facilities contracts specify requirements (e.g., a physician shall be in a plant during all three shifts).

4. Case Management Fee: The clinical case management fee is set between on-site provider and the company customer to include the initial evaluation, follow-up by the physician, and consultation by the physican throughout the term of the case. A minimum number of cases per week may be required.
5. Capitation: The method of paying for health services is based on a fixed rate per eligible member, based on benefits to be covered, not on the actual services provided to each enrollee. The provider is responsible for delivering or arranging for the delivery of all health services required by the covered person.
6. Diagnostic Related Group (DRG) Mechanism: A set number of procedures, outpatient visits, or hospital days are allowed based on a diagnostic code issued for a work-related injury. Outcome measurements drive the termination or continuation of this process.

Table 7 is an example of an industrial rehabilitation cost savings report and study conducted by the on-site contracted provider at an automobile company. This type of cost savings report shows dollar saved in travel time and mileage, number of modalities utilized, and discounted pricing.

CLAIMS AND STATUTES

Disability claims may be filed for on-the-job injuries/illnesses generally for up to one year from the time the employee sustained the accident. Employees report the incident that results in lost work time or medical expenses to the workers' compensation representative or administrator. It is important that the on-site provider understand and be familiar with workers' compensation statutes that protect both the employee and the employer. Regulatory compliance and policies, e.g., wearing of safety personal protective equipment (glasses, gloves, hard hat) in the plant and Occupational Safety Health Act (OSHA) medical surveillance standards, should be reviewed by the on-site provider. Training and education provided on site to workers reduces the frequency and severity of injury. Appendix 1 illustrates an agenda for health, safety, ergonomics, and prevention of upper extremity/hand and back injury developed and used in the training by an on-site team.

Positioning on-site occupational health and rehabilitation services in the actual work place assists companies with high-risk workers and a higher number of workers' compensation injuries/illnesses. These company customers are demanding return-to-work services as a way to reduce costs. See Chapter 5 for recommended on-site rehabilitation equipment.

External factors have created a demand for these services. On-site occupational health and rehabilitation services are increasing in popularity because they offer employers, workmen's compensation/third party payers, and insurance companies a way to reduce costs. Advancing medical technology is increasing the need for rehabilitation services by allowing victims of severe accidents to survive with physical limitations. Conditions of the aged also play an important part as the working population ages. Outcome measurements (e.g., HEDIS 2000 report card) stress the importance of diagnosing, treating, and preventing acute and chronic conditions before they become bigger problems, like a stroke or a heart attack. The Health Plan Employer Data and Information Set (HEDIS) measures health plan performance. It is compiled by the National Committee on Quality Assurance.

According to J. Duncan Moore (1999) "the two new asthma measures, for example, are designed to control the condition. If fewer plan enrollees end up in the emergency rooms, that means they're better managing their care. The proportion of enrollees with asthma who use antiinflammatory inhalers will show how well the plan is educating its patients and providers." The on-site physician provider already looks for ways to trim employer/company costs through wellness and prevention programs. As an on-site health care provider, the occupational medicine physician must become an expert in caring for illnesses or conditions and injuries blended with managed care experience. By benchmarking organized systems of care against various part of the country and tracking regional differences in care, providers can measure health costs.

Physicians and health maintenance organizations will continue to research diseases to slow the increase in health care costs. With the anticipated number of Americans age 65 or older expected to reach over 75 million in 2050, the worker and the employer must be informed and educated on site about disease prevention.

APPENDIX 1: AGENDA: HEALTH, SAFETY, ERGONOMICS, AND PREVENTION OF UPPER EXTREMITY/HAND AND BACK INJURY

		Time	Presenter
I.	Program overview (safety and injury prevention). **A.** Hand anatomy and positioning **B.** Splints **C.** Tendon glide, medial nerve glide stretching, and hand-strengthening exercises	20 minutes	On-site director
II.	Back anatomy **A.** Musculature that helps support the spine, the abdominal muscles, pelvis	15 minutes	
III.	Body mechanics/work posture **A.** Bending **B.** Lifting **C.** Pushing **D.** Pulling	12 minutes	Videotape back exercises, booklet
IV.	Posture standing/sitting demonstration **A.** Lifting table to floor (20 lbs) **B.** Overhead lift (10 lbs) **C.** Hand and pinch strength testing **D.** Body mechanics, reach, bend, twist, turn **E.** Sitting and stretch chair activity	30 minutes	Group
V.	Risk factors for back problems or injuries **A.** Lack of physical fitness (muscles, joints, etc., injure faster) **B.** Lack of dietary control (overweight, stomach protrusion)	8 minutes	

	C. Smoking (decrease oxygen in blood circulation and muscles)		
	D. Lack of flexibility (stiffness causes decreased movement)		
	E. Lack of strength (weak muscles injure easily)		
	F. Improper education on how to use body		
	G. Mental stress (inability to focus and take care of body)		
VI.	Stretching and strengthening exercises	10 minutes	
VII.	Ergonomic interventions[a]	15 minutes	
VIII.	Open discussions and questions	10 minutes	Group

[a] The term ergonomic refers to ''fitting the job to the person.'' Common ergonomic things we see in the environment include (a) hi-lo, crane, hand truck, dollies, carts, or any mechanical device decreasing our lifting, (b) drafting tabletop tilt the work toward the person rather than making the person lean over the table, (c) computers, keyboards, and stools. Ergonomics serve three primary functions: (1) increase efficiency—if something is physically easier to do it is generally more efficient (automatic systems for moving materials minimize employee handling, speeding up production and reducing physical demands), (2) increase clarify—tasks easier to understand less confusion, (3) decrease the physical demands of the task with hydraulics, etc. In closing, making changes in lifestyle to enhance wellness takes time. Habits are not changed overnight. It requires practice to change behavior.

4

Health Care Reform: Quality, Price, Service, and Outcome

Cases of patients who have sustained work-related illness and injury should be reviewed monthly by the on-site occupational health and rehabilitation team ("grand rounds"). The review should include the patient's diagnosis, prescription of treatment, quality assurance (e.g., number of visits), ancillary tests, and on-site physical and/or occupational therapy treatments. The progress of the patient in therapy will be discussed and presented by the patient's therapist. The schedule for further evaluation, treatment, and follow-up of the patient is documented with the goal of return to work in mind. Secondary diagnosis (e.g., arthritis) and systemic non–work-related patient complaints may also be discussed.

National disability management is a term frequently used to integrate non–work-related and work-related employee injuries/illnesses. For example, whether or not a worker falls off a ladder at home on a weekend or slips and falls at work on the plant floor, the issues of time off, not at work, or lost productivity still costs the employer disability dollars. Attention to prevention, case management, or grand rounds will assist companies in the medical management of claims. Early return to work pays big dividends. The same resources on site are devoted to the final outcome of cost reduction in health care, productivity, and indemnity payment. Safety and prevention as well as making recommendations for the patient to be discharged to the fitness/wellness center on site is a topic of discussion at grand rounds. Appendix 1 is a sample grand rounds/case management meeting memorandum, agenda, and several workers' compensation case studies.

Businesses should consider on-site efforts in contrast to traditional off-site medical care because companies spend billions of dollars on employees who sustain repetitive motion injuries. The on-site model is a valuable business and process management tool that supports the production and delivery of goods and services. The on-site delivery team of occupational medicine providers has a common goal with the company to optimize customer care and quality manufactured products.

The on-site occupational health staff, including physicians, physical and occupational therapists, nurses, physician assistants, and athletic trainers in manufacturing plant medical departments and other industry locations can benefit from the lessons learned in industry and manufacturing. Staff apply methods of control techniques to produce improved patient outcome. Over the past decade of collecting data where the concentration of employees make the on-site model of delivering medical care and rehabilitative services practical, the on-site model has proven to be a cost-effective alternative to sending employees off site. The on-site model reduced open workers' compensation cases (i.e., the patient returned to work or the claim was fully processed/case closed) for one corporate customer by more than 47.1% in one year alone.

The on-site, full service program is offered during all three shifts of plant operations, seven days a week. Job placement results have improved because the on-site occupational medicine physician is able to document appropriate work restrictions. Rapid response time initiating services and the ability to modify the work to accommodate restrictions knowing the limitations on the productivity of the job adds to the program's success. Clinical case management between the company/customer, the injured/ill worker, and the union is proactive. The on-site occupational medicine physican has the final say on the work status of the injured/ill employee at the plant. Education of specialists and effective communication between the on-site occupational medicine physician and the off-site specialist (e.g., hand surgeon) is essential. In order to make this venture a win-win situation it is important to work together.

The occupational therapist is an expert in functional capacity evaluation and ergonomics. The therapist guides the injured/ill worker through the work restriction physical exertion levels on site in preparation for return to work at the preinjury level. Pain management is provided through the industrial rehabilitation program. The U.S. Labor Department's proposed ''ergonomics rule'' would require employers to correct conditions that cause cumulative injuries. The on-site model is preventative. Employees are educated to use the ergonomic equipment properly. Training programs on site teach workers how to prevent muscle strain and to use correct body postures and body mechanics.

There are so many variables in worker size, height, coordination, strength, and range of motion that an ergonomist cannot invent a production line that will accommodate all body sizes. The U.S. Labor Department estimates that 1.8 million workers annually have work-related musculoskeletal injuries and that 600,000 people miss work because of these injuries, costing $15–20 billion annually in workers' compensation and $30–40 billion in other expenses, such as medical care. The injuries to muscles, nerves, ligament and tendons include such problems as carpal tunnel syndrome, back pain, and tendinitis.

The ergonomics rules would cover a broad range of workers, from nurses' aides who must lift heavy patients to baggage handlers at airports to people who work on computers or assembly lines. OSHA estimates the rules could prevent injury to 300,000 workers annually and save employers $9 billion.

Under the proposed rules, a worker who has an ergonomic injury diagnosed by a doctor would be entitled to have the work environment fixed to relieve the cause (e.g., by changing the height of an assembly line or computer keyboard). A worker who must be assigned to lighter duty during recovery from ergonomic injury would be guaranteed normal pay and benefits. A worker who must leave the job altogether would be guaranteed 90% pay and full benefits during recovery. At workplaces with numerous incidents of ergonomic injury, employers would have to provide medical help and safety retraining for workers in addition to fixing physical problems. The rule also would require all manufacturers and companies with workers who do manual heavy lifting to provide preventive training. Proposed new regulations will continue to force employers to engage in proactive strategies to control costs. The on-site model is uniquely positioned to provide services to employers in a cost-effective manner.

Documentation of an anticipated date for the workers' recovery is performed by an occupational medicine physician who is familiar with the workstation, space, tool use, production layout, and speed of production. Disability management on site allows for reviewing the compensability of the claim and ensuring that the accident has been investigated. Maintaining contact with employees with work-related injuries or illnesses and acting as an advocate for employees when appropriate is included in the medical care on site.

The on-site program outcome resultsd are shared with the employers. The following areas have been evaluated and shown to benefit the employer as well as the employee utilizing the on-site model of delivery:

1. Decrease Employer Indemnity Costs—*Compare*:
   Lost work days
   Restricted work days
2. Reduce the Frequency and Severity of Work Injury—*Measure*:
   Number of OSHA recordables
   Visits to plant medical facilities
3. Close Workers' Compensation Cases—*Record*:
   Cost of litigation (cases settled and legal fees)
   Return-to-work-rates (to original job; permanent disability; limited/ light-duty placement; restrictions eliminated)

4. Decrease Medical Expenses—*Track*:
   Outside referrals to specialists
   Reduction in reserves (funding to pay estimated liability)
   Travel time (i.e., productivity), mileage, and number of modalities, procedures, tests
5. Increase Patient/Employer Satisfaction—*Record*:
   Function and productivity rate
   Number of plant ergonomic modifications

Providing health care and follow-up on site is an avenue for setting up procedures for workers to report symptoms. Educating workers as to risk factors and signs and symptoms of musculoskeletal trauma in the workplace can reduce work injury and hazards. Progress can be tracked by plant medical employees. Protection of the worker from repetitive-stress injuries is cost effective. The on-site model has been shown to decrease costs by demonstrating the dollar amounts associated with lost work time and medical expenditures.

Education is essential for quality process improvement. Meeting and exceeding service standards using quality and quantity measurement tools and documentation of attendance by staff at these programs is key. Appendix 2 shows an example of an on-site staff inservice meeting minutes.

As noted in the inservice meeting minutes (Appendix 2), the Lumbar MedX Machine is present on site in three of the on-site rehabilitation company plants and has been utilized to benchmark outcomes. The outcome is measured by the functional activity of returning back to work. Our successes with the workers' compensation patient have included a 74–80% return to work. Physical therapists also utilize other modalities and treatments such as the McKenzie, body mechanics, soft tissue mobilization, myofascial release, trigger point release, and other muscularskeletal techniques in conjunction with the MedX machine. Although weaker workers may be at greater risk for injury to the back while performing a full body lift (as in lifting a tire), there has been no lifting strength correlation to back injury studied on site at our clinics. The physical demands of the job are classified by the registered occupational therapist during on-site job assessment. General aerobic fitness alone, however, seems to have no relationship to workplace injuries, as shown in the Boeing study (Battie et al., 1989). Mooney et al. (1995) studied workers in the coal strip-mining business, where the incidence of torso injuries was 63% and strains to the low back was 59%. The purpose of their study was to demonstrate the effect of a once-a-week exercise program focused especially at lumbar extensor strengthening. Low back claims studied by the same group involved 197 male workers who volunteered to exercise compared to workers who did not exercise. All participants were placed into the same MedX exercise protocol for 20 weeks. There was a 54–104% increase in strength during a 20-week program (Mooney et al., 1995).

The MedX lumbar extension machine is one tool used to treat backs. The MedX lumbar extension machine outcome reports were collected on site at an automobile assembly plant. Physical therapists were provided on site for 12 months. Of 124 patients, 31, or 25%, of the back cases were treated with the MedX lumbar extension machine. The MedX protocol was followed and involved progressive, resistive exercises of the isolated lumbar spine with the pelvis firmly stabilized. All patients also completed a generalized physical therapy treatment program. Back diagnosis seen during the year included chronic low back pain, acute lumbar strain, status post–Herrington Rod, sacroiliac dysfunction, herniated lumbar disc, lumbar myositis, lumbar degenerative disc disease, status post–lumbar laminectomy, low back strain, lumbar spondylosis, L5 radiculopathy, lumbar contusion, acute sciatica, sacroilitis, and acute lumbar spasm.

Based on 124 workers' compensation discharges for back cases, 29 or 23.39% remained on sick leave after discharge. Ninety-five or 76.61% were physically able to work or continue working. The MedX results for the 25% of all lumbar cases placed on the MedX were as follows:

8 of 31, or 25.8%, remained on sick leave after discharge
23 of 31, or 74.2%, were physically able to return to work or continue working

Physical therapy results without MedX were as follows:

21 of 93, or 22.6%, remained on sick leave after discharge
72 of 93, or 77.4%, were physically able to return to work or continue working.

The sample reported is small in comparison to the 627 chronic low back pain cases who completed a MedX program and in which at one year follow-up 94% of patients with good or excellent results reported maintaining their improvement (Nelson et al., 1995).

Manufacturing plants vary in the degree of repetitive work required. The speed of the assembly process in an automobile assembly plant is much greater than for the average bank teller, nurse, or nonproduction line worker. The return-to-work rate for any diagnosis needs to be viewed in terms of the physical demands of the job.

Many steel and manufacturing plants enable workers to periodically transfer among a number of different positions and tasks within jobs that require different skills, postures, and responsibilities. Rotating job tasks assists workers understand the different steps that go into creating a product or delivering a service. Cross-training can also contribute to productivity by helping workers to understand the job and its physical demands. The workers must possess a general knowledge of the skills required to perform a variety of work functions. Job rotation has been utilized as a way for injured/ill employees to identify for the on-site medical team which jobs in the plant the patient/employee thinks he or she can perform within a particular work restriction. It allows for increased motivation and participation by the patient/worker in placement and work conditioning recovery processes.

Disability management and any national health care reform policy is focused on quality of care, informed consent, early intervention, low cost, and the benefits of ordering testing and procedures. Little attention is paid to the patients themselves, who ultimately demand choice and quality. Productivity and earnings motivates the workforce. Therefore, surveying patients as to customer satisfaction will benefit the company, and the on-site provider. Job retention and seniority on the job are the motivators for the on-site provider team to keep injured/ill workers satisfied. Knowledge that jobs are available or can be identified as restricted jobs will allow patients to gradually return to their regular duties. Appendix 3 is a sample patient satisfaction survey collected over a 6-month time period. It is an example of a tool utilized to measure customer satisfaction at an on-site industrial rehabilitation facility. Appendix 4 is the actual survey tool questionnaire, completed by the patient on his/her last day of treatment in the industrial rehabilitation clinic, and Appendix 5 is a sample of an occupational medicine physician and nursing staff patient satisfaction questionnaire.

As discussed in Chapter 2 (Appendix 3), a joint on-site industrial rehabilitation team and labor and management advisory board meets quarterly to evaluate clinical medical record quality. The advisory board members make their interest known to further discuss continuous quality improvement, policy and procedure, and research, study, and survey data.

The daily goal of the on-site occupational health and industrial rehabilitation team is to reduce workplace injuries/illnesses, to enable injured/ill employees to continue working, and to minimize lost work time of injured/ill employees. Keeping these goals foremost in mind when delivering workers' compensation care, the practitioner has a responsibility to reform the way that health care is traditionally administered and focus on lowering health care costs and improving quality. This theme is expressed by many employers as well. This has also been noted by large manufacturing companies in the media. The greater the data sharing with employers and benchmarking between business, insurers, and health care providers, the less variation will occur, making possible lower cost and higher quality.

Concerns regarding the use of certain unlicensed personnel to keep costs down raises fundamental concerns and should be given serious consideration. The education and clinical backgrounds of physical therapists and kinesiotherapists are not comparable and the services furnished by each is in no way interchangeable. Physical therapists attain their prerequisite skills through extensive academic and clinical education. The minimum educational requirements for a physical therapists is a 4-year college degree in physical therapy from an accredited educational program (American Physical Therapy Association, 1995). Physical therapists are licensed health care professionals, as are occupational therapists. In the best interest of patients and rehabilitation staff, it is crucial to be cognizant of the distinct role each health occupation plays within the health care delivery system. Kinesiotherapists are not a substitute for a physical therapist or an occupational therapist. Athletic trainers are also distinct. Improper utilization of personnel can confuse, mislead, or, worse, jeopardize the health and well-being of the consumer. In the interests of each profession, the term ''physical rehabilitation'' cannot be substituted for ''physical therapy.''

The Council on Professional Standards for Kinesiotherapy and the American Kinesiotherapy Association offer information on the role and scope of practice, as does The American Occupational Therapy Association, and The American Physical Therapy Association. Quality and low cost must be measured by outcome. Cuts in reimbursement and consumer choice is causing the health care provider to evaluate spending and quality. A recent American Medical Association survey ranked quality of health care fourth in importance behind cost, access, and managed care in the opinion of both the public and physicians (American Medical Association, 1997a,b).

To be useful, the tools and/or methods of quality measurement should be responsive to the needs of purchasers, health care providers, professionals, organizations, and patients. Information systems and technologies can assist recipients and providers through telemedicine education programs at a distance, through standardized computerized medical records, and through internet diagnostic protocols and data collection. Consumer choice is dependent on large groups and organization of purchasers who have knowledge of provider data demonstrating the tracking of processes and outcomes of care. The rates of provision of prevention services and medical record chart audits provide good comparative data, as shown in Figure 1.

Outcome management in health care is geared toward methods, applications, and results. Flowcharts of clinical processes and patient outcome focus on providing benchmarking. Statistical quality control to systematically assess and track data is completed every quarter. To consistently document and to prevent variation in treatment patterns, the number of treatments provided to back and carpal tunnel syndrome patients by the physical and occupational therapists are recorded (see Figs. 2 and 3). The ability to link clinical and financial information to assess resource use in the health care–delivery process is documented to assure quality.

The on-site occupational health and industrial rehabilitation team utilizes a customer feedback process form (see Appendix 6) as a quality assurance worksheet for a peer review of active and retroactive patient medical records. This simplifies compliance activities and improves on organization clinical performance.

Report cards on health plans, hospitals, and medical groups have appeared on the front pages of newspapers, on television, and on the internet with projects

Quality Assurance Report
Medical Record Chart Audit
Quarter Ending 6/30/99

On-site Industrial Rehabilitation Facility—2nd quarter
Reviewer: Dan Stone, A.C.S.W. Threshold = 95% compliance
Charts Reviewed = 27 Active = 14 Discharged = 13

	Quarter Comparison	
Findings of Compliance	1st	2nd
1. M.D. orders		
Duration of treatment = 53	91%	100%
2. Initial evaluation		
a. Measurable treatment goals—27	96%	100%
b. Treatment plan—27	100%	100%
c. Rehab potential—27	100%	100%
d. Frequency and duration of treatment—27	100%	100%
3. Documentation that patient is seen by a physician every 30 days—27	100%	100%
4. Daily notes		
a. Completed after every treatment—320	99%	100%
5. Discharge summary		
a. Record of progress—13	100%	100%
b. Discharge notes filed within 2 working days—13	100%	100%

FIGURE 1 Sample quality assurance report—on-site industrial rehabilitation facility.

to increase quality. Academic medicine, businesses, and the government describe strategies to evaluate overuse to underuse of effective medical procedures and services. National guidelines have been written. Measuring quality has been defined by the Institute of Medicine as ''the degree to which health services for individuals and populations increase the likelihood of desired health outcomes and are consistent with current professional knowledge (Chassin and Glavin, 1998).

An informational think tank of employer organizations has placed clauses in contracts with health plans that require specific improvement in quality. Quality is usually measured by reporting achievable goals with each intervention, by processes of care, by percentage of patients undergoing a particular examination, or by time spent waiting for appointments. Age, severity of illness or injury, sex,

Quality Assurance Report
Medical Record Chart Audit
Quarter Ending 3/31/99

Hospital Setting Facility—1st quarter Threshold = 95% compliance
Reviewer: Dan Stone, A.C.S.W.
Charts Reviewed = 69 General = 27 Back-specific = 42
General: Active = 13 Discharged = 14

Findings of Compliance	Quarter Comparison 4th	1st
1. Physician Orders		
a. Duration of treatment—59	95%	98%
b. Frequency of treatment—59	95%	98%
c. Recertification every 30 days—33	75%	96%
2. Initial evaluation		
a. Measurable treatment goals—27	95%	96%
b. Treatment plan—27	95%	96%

Clinical Review—Number of Visits in Physical Therapy

I. Back review of:
52 patients with back diagnosis were seen in the first 6 months, although only 42 were reviewed due to incomplete documentation to contrast goals from the initial evaluation to the discharge summary. Since many patients were still active in therapy, the discharge summary was not documented.

Diagnosis:
- Lumbar radiculopathy—18
- Low back pain—7
- Lumbar disc herniation—6 (includes one postsurgical)
- Lumbar strain—5
- Lumbar syndrome—2
- Sciatica—2
- Spondylosis—1
- T5-T6 compression fracture—1

Acute = 27 Chronic = 15

Treatment ranged from 3 to 52 treatments, with the average being 11.

Goals met:
49% and below = 9
50–74% = 6
75–100% = 27

Reason for goals not met:
1. Noncompliance with physical therapy = 5
2. Patient was compliant but didn't respond to therapy = 14

FIGURE 2 Sample quality assurance report—hospital setting facility.

II. Percentage of canceled/no-show appointments. The cancellation/no-show rate for the on-site industrial rehabilitation site was 12.5% of total appointments.

Note: The most significant change in results was the increase from 15% to 96% this quarter in recertification of physician prescriptions every 30 days. Suggestions were presented to monitor the transition from the rehabilitation center to the wellness/fitness center through therapist documentation. Patients are recommended to exercise postdischarge from therapy in the fitness center. Physical and occupational therapists show patients the correct way to exercise on the fitness machines to prevent reinjury. Data will be collected to compare rate of reinjury of patients exercising in the wellness/fitness center post–industrial rehabilitation.

FIGURE 2 Continued

race, health plan, geographic access, and choice are all factors utilized to measure quality. Evidence-based hospital referral, involves the channeling of patients to certain hospitals for certain conditions and procedures (including coronary angioplasty and bypass surgery, carotid endarterectomy, and repair of abdominal aortic aneurysm) for which clear evidence exists that a higher volume of procedures or teaching status is associated with better outcomes (Hannan et al., 1992; Jollis et al., 1994; Grumbach et al., 1995; Karp et al., 1998). Some companies reduce premiums for employees who choose high-quality plans (Bodenheimer, 1999).

Patient safety is one of the targeted issues. Education will advance evidence-based hospital referral and reduce medication errors through computerized

Quality Assurance Report
Quarter Ending 9/30/99

Clinical Chart Review—Number of Visits in Occupational Therapy
Repetitive motion diagnosis—6
Diagnosis: carpel tunnel syndrome—3 (2 bilaterally affected)
DeQuervains—3 (includes DeQuervains tenosynovitis and DeQuervains/reflex sympathetic dystrophy dual diagnosis)
Treatment range from 6 to 18 treatments, with average being 7
Goals met—49% and below = 0
50–74% = 1
75–100% = 5

FIGURE 3 Sample quality assurance report—occupational therapy visits for repetitive motion disorders.

physician order entry systems displaying warning of drug interactions. The on-site model of occupational health and rehabilitation is best operated under a contractual relationship where it is part of a large, vertically integrated health care organization that is at the forefront of new technology offering advanced treatments.

The on-site model that is connected to the use of many tools to measure performance and to chart such different factors as success rates helps to balance priorities, and it links up to system-wide indicators in clinical practice. The worker on site is well managed, which can positively impact quality of care and costs. Whether the worker has occupational asthma, chest pain, diabetes, or other chronic conditions, the on-site clinical team should have access to proven best practices and approaches. The injured/ill worker is case managed throughout the episode of care by the on-site team.

In 1993 President Clinton's proposed plan for universal health insurance was introduced, but, it was defeated by opponents who wanted to package in a better way. National disability management, commission on health care reform, and universal care are all terms used to guarantee access of Americans to medical care. The on-site model responds to the community, patient, and health system's objectives of quality, cost, convenience, and service. The employee working until age 67 is expected to be an active consumer and to be a full participant in health care choice.

The on-site model of occupational health and rehabilitation that is paid for by the employer at a monthly management rate offers the employee/worker access to health care at affordable rates. The on-site physician goals are to establish a rapport with the worker and makes suggestions for behavioral modification of lifestyles (e.g., diet, exercise, and preventive services), to enhance patient safety and expand care, to slash today's wasteful overhead typical of some insurance bureaucracies, and to, in the long run, improve workers' health and productivity. The increased availability of services provided on site through the occupational health and rehabilitation model will for the next 10 years, and especially the next 5, be largely obtained with prevention and health care advancements offered to potential new customers.

Customers (i.e., company/employers) need to be screened through a series of interviews before and after they choose the on-site model of care. Loyal customers want to work with suppliers of health care. At first glance a customer may be valuable, but the "contribution margin," or what is left over to help cover the health care provider's overhead and contribute to profits, must be considered. Understanding the employer's culture means becoming familiar with the philosophy of the plant operations, production system, and competitive manufacturing as well as the human resources policies and operating principles. The contract for on-site occupational health and rehabilitation has volatility. Such a contract is usually awarded for no more than 3 years.

APPENDIX 1: SAMPLE OF GRAND ROUNDS MEETING, AGENDA, AND CASES DISCUSSED

Memorandum

TO: Safety Coordinator
Physical Therapist
Occupational Therapist
Staff Plant Occupational Medicine Physician
Office Coordinator
Certified Occupational Health Nurse
Human Resources Director
Workers' Compensation Representative
Claims Administrator
Rehabilitation Technician
R.N. Case Manager
Job Placement Manager
FROM: Administrative Director, Department of Occupational Health
RE: Grand Rounds/Case Management Meeting
DATE: March 10, 1999

The March 1999 Patient Case Study/Grand Rounds meeting is scheduled for Wednesday, March 10, 1999, at 1:00 p.m. Attached is a copy of the agenda with a listing of the patient names whose cases will be discussed.

Anticipated date of recovery, progress in physical therapy and occupational therapy, and ergonomics will be discussed. Light-duty transitional work placement or return to the same job assignment will be discussed concerning each patient case.

I look forward to seeing you. If there are any patients that you would like to discuss who are not on the list, please feel free to add them during the meeting.

If you have any questions regarding the attached agenda, please contact me at (phone number). Thank you.

Grand Rounds Meeting Agenda, Wednesday, March 10, 1999

Grand Rounds, Patient Case Review and Management 2 hours

	Patient Name	Work Location	
1.	Joe Thomas	Woodshop/ Metal Model Maker	10 minutes
2.	Debbie Smith	Housekeeping/Janitor	5 minutes
3.	John Dow	Machine Operator	10 minutes
4.	Bob Banks	Electrician	10 minutes
5.	Cheryl Flower	Transmission Building	10 minutes

Usually 14–17 cases are reviewed within a 2-hour period.

The following are examples of patient cases discussed by the doctors and physical/occupational therapists and other grand rounds team members.

1. Diagnosis: lumbar radiculopathy. Patient is a 44-year-old metal model maker. Patient started physical therapy in February. Patient continues to have a dysfunction with his bladder and has had to go in for emergency surgery. He missed 3 days of work and had the weekend to recover. Patient has had 2 weeks of physical therapy, four treatments so far. Patient is able to get up on his heels and toes now. He still continues working in the plant as a metal model maker, on day shift, Monday through Friday. He works from 6:00 a.m. until 2:30 p.m. Patient is improving. Patient has a foot drop. Patient is working after therapy. He attends therapy for 1 hour (8:00 a.m. to 9:00 a.m.). Recommendation is to continue physical therapy for 1 more week. Patient may need an orthotic device.
2. Diagnosis: reflex sympathetic dystrophy (RSD). Patient had blood work done and the results came up negative for connective tissues disease. Patient is improving. She wears a right-hand jobst glove every day but takes it off when sleeping. She is right hand dominant. Patient has not missed work in housekeeping while she has attended occupational therapy. She needs to work on strengthening, tendon glide, edema control, and weight-bearing and stress-loading exercises. Recommendation is to continue occupational therapy three times a week for 3 weeks and to increase her to 62 pounds of grip strength in six visits, then discharge. The RSD protocol will continue. Ergonomic

techniques and tools were discussed with patient (i.e., scrubber with automatic load for soap). She is 5 months status post right colles fracture.

3. Diagnosis: right index finger amputation, proximal interphalangeal joint is preserved. Patient has had eight occupational therapy sessions so far. Patient is compliant with treatment. January 24th is the injury date. Dr. Ditmers is the physician who did the surgery. Patient is a machine operator, he is right hand dominant. He continues working in transitional work consisting of dispatch in the security department. He only missed 5 days of work from this injury. Status post history of left index finger crush injury as well. Therefore, bilateral hand grip strength is weak. During reevaluation he was at 60 degrees of flexion at the right proximal interphalangeal joint; patient has gained in active range of motion by 15 degrees. Patient is very motivated. Recommendation is to continue occupational therapy with rehabilitation goals of increasing active range of motion strength, a desensitization program, and edema control for return to work as a machine operator with four more visits in hand therapy. The right third digit will be utilized with the thumb for pushing buttons. The physician follow-up report stated good bone and tendon integrity. Return to work at full capacity to original job in 2 weeks. Home program can be demonstrated by the patient. He is motivated and compliant with placement, therapy, and medical staff.

Respectfully submitted,

____

Business Office Coordinator

APPENDIX 2: INSERVICE/STAFF MEETING MINUTES

TIME: April 28, 1999
TIME/PLACE: 1:00 p.m./Conference Room 1-A
SUBJECT: Inservice on the MedX Lumbar Extension Machine and the research behind the equipment's success.

All personnel are required to sign your name in and out for proof of attendance at the inservice meetings. A sign-in sheet has been provided for signature. Attendance is mandatory. This record will be kept on file to verify attendance requirements.

The University of Florida, College of Medicine, College of Health and Human Performance has been instrumental in testing the MedX spine protocols. MedX certification and demonstration of the MedX lumbar machine was discussed by Jim Flanagan and Robert Secora. The founder and former chairman of Nautilus Sports Medical/Industries and MedX Corporation, key note speaker Arthur Jones discussed the history of the development of MedX. Rotary movement that is limited and at controlled speed of movement is the rehabilitation principle. Pelvic stabilization and gravity compensation occurs. A full range-of-motion evaluation is completed. Dr. Fulton at the University of Florida has studied 300 patients with low back pain and sciatica, back pain without sciatica, myofascial syndrome, spinal stenosis, lumbar spondylosis, and lumbar instability. Chronic back pain patients are measured with outcome presented by functional activity. Bone mineral density changes have been positively noted in 40 patients.

The MedX treatment combines the forces of activity, structural integrity and injury for meeting the goal of a strong back with increased range of motion. Post treatment on the MedX the patient is encouraged to come back to the clinic once a month to maintain the same intensity as at discharge from therapy. Nationally and locally benchmarking against plants with MedX, our own on-site rehabilitation team has achieved a 74–80% success rate of returning patients back to work post MedX treatment. Patients also report increased ability to perform activities of daily living. The MedX machine is utilized along with other physical therapy modalities and treatments including the McKenzie, body mechanics, and soft tissue mobilization. In some settings the MedX has been used for preplacement strength testing.

Staff asked questions and the meeting adjourned at 2:15 p.m.

Respectfully submitted,
Jane DeHart, MA, OTR
Administrative Director

APPENDIX 3: SAMPLE AUTOMOTIVE INDUSTRY ON-SITE REHABILITATION PATIENT SATISFACTION QUESTIONNAIRE

I. Background

The on-site industrial rehabilitation provider team felt it was important to determine patient satisfaction levels related to their facility location. There was an interest in outlining, under confidential and anonymous conditions, the distance the patient travels, who chose the location for treatment, satisfaction levels related to various treatment aspects, waiting list length, waiting time for scheduled appointments, home program, wellness/fitness center attendance, and any additional comments the patient has regarding the program. These questionnaires were gathered during the period covering June through December, 1998.

II. Purpose of Research

The staff at the on-site industrial rehabilitation facility were interested in patient perceptions regarding the facility and the treatment they received during their visits. It was important to determine if patients had been referred to the correct facility as well as any areas of services that the patient indicated were weak or was dissatisfied with. Asking for patients' comments and perceptions regarding the service aspects allows the staff to change the weak service areas and to provide better treatment and/or services to present and future patients. These improvements will benefit not only patients but may generate additional business for the rehabilitation facility.

III. Survey Limitations

The surveys were administered to patients at the end of their final treatment session at the on-site industrial rehabilitation facility. Patients may feel pressured to answer the questions in a positive manner due to the setting, which would cause the survey results to become skewed in a positive manner (i.e., not adequately reflecting problems or difficulties the patient perceived or experienced during their time at the facility).

IV. Project Objectives

1. To evaluate patient responses to facility comfort, cleanliness, and appearance.
2. To evaluate patient-staff interactions in terms of time spent by the therapist with patients and perceived quality of care provided by the therapist and the therapist's availability to answer the patient's questions.
3. To determine who decides where treatment will take place.
4. To review patient scheduling practices within the facility locations to determine if patients are having to wait too long to get an appointment or if the appointments are being scheduled without delaying the next patient.
5. To determine if the patient is going to continue exercising in the wellness/fitness center.
6. To consider patient's responses and comments in making changes in the program that will benefit both patients and staff members in the areas of treatment quality, time availability, or scheduling format.

VI. Survey Questionnaire Results

For the on-site industrial rehabilitation facility survey, questionnaires were gathered for the period covering June through December, 1998. There were 57 surveys collected during this period. This is the 20th group of surveys collected from the on-site industrial rehabilitation facility. The original group of surveys were gathered during the first quarter of 1994. During this 1994 period 8 surveys were collected. The second group of surveys was collected from April through July of 1994, with 32 patients surveyed. The third group of surveys was gathered during the third and fourth quarter of 1994. During this period 22 surveys were collected. From January through March of 1996 a total of 42 surveys was collected.

Below are the raw scores and the percentages for each question asked on the survey questionnaire.

Question 1: The distance traveled for treatment.

Distance	No. of Responses	% of Total
0–5 miles	47	82.5
6–10 miles	4	7.0
11–15 miles	4	7.0
16–20 miles	0	0.0
Over 20 miles	2	3.5
Total	57	

Question 2: Who chose the treatment location?

Decision Maker	No. of Responses	% of Total
You (patient)	28	49.0
Your physician	13	23.0
Your family	2	3.5
You and physician	8	14.0
Other—no response	6	10.5
Total	57	

Question 3: Rating of treatment aspects/program services.
A. Quality of care by the therapist.

Ranking Number	Rating Level	No. of Responses	% of Responses
1	Very dissatisfied	0	
2	Dissatisfied	0	
3	Neutral	0	
4	Satisfied	8	14
5	Very satisfied	49	86
	Total	57	

B. Time spent by the therapist.

Ranking Number	Rating Level	No. of Responses	% of Responses
1	Very dissatisfied	0	
2	Dissatisfied	0	
3	Neutral	0	
4	Satisfied	15	27
5	Very satisfied	42	73
	Total	57	

C. Availability of the therapist to answer questions.

Ranking Number	Rating Level	No. of Responses	% of Responses
1	Very dissatisfied	0	
2	Dissatisfied	0	
3	Neutral	0	
4	Satisfied	8	14
5	Very satisfied	49	86
	Total	57	

D. Courtesy and friendliness of nonmedical staff.

Ranking Number	Rating Level	No. of Responses	% of Responses
1	Very dissatisfied	0	
2	Dissatisfied	0	
3	Neutral	0	
4	Satisfied	7	13
5	Very satisfied	50	87
	Total	57	

E. Comfort of the facility.

Ranking Number	Rating Level	No. of Responses	% of Responses
1	Very dissatisfied	0	
2	Dissatisfied	0	
3	Neutral	0	
4	Satisfied	0	
5	Very satisfied	57	100
	Total	57	

F. Facility cleanliness and appearance.

Ranking Number	Rating Level	No. of Responses	% of Responses
1	Very dissatisfied	0	
2	Dissatisfied	0	
3	Neutral	0	
4	Satisfied	0	
5	Very satisfied	57	100
	Total	57	

G. Was the phone staff responsive to your needs?

Ranking Number	Rating Level	No. of Responses	% of Responses
1	Very dissatisfied	0	
2	Dissatisfied	0	
3	Neutral	0	
4	Satisfied	0	
5	Very satisfied	57	100
	Total	57	

H. Patient's overall satisfaction with on-site rehabilitation.

Ranking Number	Rating Level	No. of Responses	% of Responses
1	Very dissatisfied	0	
2	Dissatisfied	0	
3	Neutral	0	
4	Satisfied	9	16
5	Very satisfied	48	84
	Total	57	

Question 4: Waiting period for an appointment.

Time Period	No. of Responses	% of Responses
0–2 days	52	91
3–5 days	5	9
5–7 days	0	
7+ days	0	
Total	57	

Question 5: Waiting period before scheduled appointment.

Time Period	No. of Responses	% of Responses
0–10 minutes	57	100%
15–20 minutes	0	
20–30 minutes	0	
30+ minutes	0	
Total	57	

Question 6: Did you receive a program to do at home?

Response	No. of Responses	% of Responses
Yes	53	93
No	4	7
Total	57	

VII. Patient Comments

Question 7: Things you like or dislike about on-site industrial rehabilitation:

a. The people here are nice.
b. Everything is fine here. I'm in light-duty work with no use of right hand for 2 weeks. My hand therapist increased my grip strength.
c. The place is real convenient with my job.
d. Although it has been 3 months since my original injury, the therapy I received here did help!!
e. I couldn't find any relief for pain—heavy pain relievers weren't even

working. Five visits here, I feel a far sight better. God bless every one of you.

f. Tess, Tony, and Deb are very nice people to deal with. Jane should be proud to have people like them working for her.
g. I feel like on most Fridays they are very busy. Even though they were busy I was able to get in and out in an hour and 15 minutes. The tech. that does the traction is very knowledgeable. They all do a wonderful job.
h. Very friendly staff. I went back to work building transmissions after physical therapy and occupational therapy treatment of eight visits.
i. Very friendly, knowledgeable, and helpful.
j. Everything was excellent. No complaints about anything or anyone.
k. The staff and my therapists ''Tess'' and ''Tony'' have helped me emotionally and physically to work on myself and hopefully prevent surgery.
l. The location is very convenient for me being that I work approximately 1/4 mile away. The staff was very courteous and helpful, my pain decreased too.
m. They are very compassionate and always ask me how I was feeling. If I have to have any more treatments I will come here!!
n. I was very pleased with everyone and the treatment equipment. The trigger points, mobilization, and stretching helped me.
o. All the people were nice.
p. Everyone was very helpful and congenial.

Question 8: Please comment on any other aspects of the services you received or on any ideas that you may have for improving services on-site.

a. Thank you very much. I would not hesitate to return if needed.
b. Their expertise and up to date skills were great, and I like that!
c. My legs feel stronger than they have in years. Both legs feel better than they did prior to my knee injury.
d. Possibly be able to use the wellness/fitness center.
e. Music would be nice.
f. I've had other therapy at a local hospital and had no results with my therapy. I came here and could see a big change in just 3 weeks with the left arm, which was very painful for at least a year.
g. I was glad that I didn't have to drive so far for quality treatment. I want to thank all the nice people for all they have done for me.
h. I feel the staff here is doing a fine job. Tess, Tony, and Deb always seem to give 100%, that's nice to have when you're not feeling well.
i. Tess—she is outstanding. The best!

VIII. Summary/Conclusions

Based on the survey information, the patients at the on-site industrial rehabilitation industry facility are overall very pleased with the staff members and the program services. Patients are generally satisfied with all aspects of this facility and felt improvement in their condition.

APPENDIX 4: ON-SITE INDUSTRIAL REHABILITATION PATIENT SATISFACTION QUESTIONNAIRE

You are an important person and helping you is important to us. To help us continue to improve our services, please answer the following questions.

THIS SURVEY IS CONFIDENTIAL. YOU DO NOT HAVE TO SIGN YOUR NAME.

When you have finished the survey, fold and place in the envelope at the front receptionist desk.

Thank you for your cooperation and for allowing the on-site industrial rehabilitation team serve your rehabilitation needs.

1. Approximately how many miles did you travel to receive rehabilitation treatment?
   __ 0–5 __ 6–10 __ 11–15 __ 16–20 __ over 20
2. Who chose this location for your treatment?
   __ you __ your doctor __ your family
   __ you and your doctor together __ other
3. How satisfied were you with each of the following aspects of your treatment?

	Very Dissatisfied	Dissatisfied	Neutral	Satisfied	Very Satisfied
a. Overall quality of care provided by therapist	1	2	3	4	5
b. Amount of time spent by therapist	1	2	3	4	5
c. Availability of the therapist to answer questions	1	2	3	4	5

	Very Dissatisfied	Dissatisfied	Neutral	Satisfied	Very Satisfied
d. Courtesy and friendliness shown by nonmedical staff members (for example, the receptionist)	1	2	3	4	5
e. Overall comfort of facility	1	2	3	4	5
f. Cleanliness and appearance of facility	1	2	3	4	5
g. When calling by phone, our response to your needs	1	2	3	4	5
h. Overall, how satisfied were you with the care you received at the on-site industrial rehabilitation facility?	1	2	3	4	5

4. How long did you have to wait for an appointment to get into your rehabilitation program?
   __ 0–2 days __ 3–5 days __ 1 week __ over 1 week
5. How long did you have to wait for your scheduled appointments?
   __ 10 min. __ 15–20 min. __ 1/2 hour __ more
6. Did you receive a home program that you can demonstrate?
   __ Yes __ No

Further Comments

7. Things you like or dislike about the on-site industrial rehabilitation:
   ____
   ____
8. Please comment on any other aspects of the services you received or on any ideas that you may have for improving services at the on-site industrial rehabilitation facility:

APPENDIX 5: CUSTOMER SATISFACTION QUESTIONNAIRE: DEPARTMENT OF OCCUPATIONAL HEALTH

1. Date: ________________________________
2. Time arrived: ________________________________
3. Company by which you are employed: ________________
4. Service required: (Please check one)
   __ Workers' compensation claim (initial first time visit for incident)
   __ Follow-up examination (workmen's compensation claim)
   __ Preplacement/new hire physical
   __ Return-to-work physical (not related to workers' compensation claim)
   __ Other
5. Did you have difficulty scheduling your appointment? Yes __ No __
6. Total time spent in waiting room: (please check one)
   5 minutes __ 10 minutes __ 1/2 hour __ 45 minutes __
   1 hr __ 2 hrs __
7. I was received at the front desk in a courteous and professional manner:
   Strongly Agree __ Agree __ Disagree __
   Strongly Disagree __ N/A __
8. The nursing staff treated me in a courteous and professional manner:
   Strongly Agree __ Agree __ Disagree __
   Strongly Disagree __ N/A __
9. The physician was knowledgeable and courteous:
   Strongly Agree __ Agree __ Disagree __
   Strongly Disagree __ N/A __
10. The staff took sufficient time to answer my questions and explain my condition:
    Strongly Agree __ Agree __ Disagree __
    Strongly Disagree __ N/A __

11. The examination facilities were adequate and allowed sufficient privacy:
    Strongly Agree ___ Agree ___ Disagree ___
    Strongly Disagree ___ N/A ___
12. Total time spent in occupational health department: ___________

Personal comments: ____

____

____

____

APPENDIX 6: CUSTOMER FEEDBACK PROCESS FORM

Quality Assurance Worksheet—Industrial Rehabilitation

Medical Record Chart Review

Name: ______________________________
Date: ______________________________
Reviewer's Signature: ______________________________

1. Identifying Information
   A. Patient name and address
   B. Patient insurance information
   C. Patient date of birth
   D. Facility name and address
   E. Physician name and address
   F. Patient work and job title history
2. Medical Record
   A. Patient's diagnosis
   B. Physician's orders
   C. Information release signed
   D. Work restrictions
   E. Appropriate physical therapy, occupational therapy
   F. Contraindications
3. Prescription/Physician's Orders
   A. Date of order
   B. Diagnosis(es)
   C. Type of treatment(s)
   D. Body area(s) to be treated
   E. Frequency of treatment
   F. Duration of treatment
   G. Changes in treatment plan or orders to continue treatment
   H. Physician's signature
   I. Recertification every 30 days
4. Initial Evaluation
   A. Date of evaluation
   B. Primary and all pertinent secondary *diagnoses* with date(s) of onset (recognized diagnoses, *not* symptoms)

C. Prior hospitalizations and surgeries *with dates*
D. Other relevant patient history (e.g., exacerbation of a chronic illness, accidental injury, past treatment received, etc.)
E. Physical complaints stated and medications
F. Contraindications, precautions if any
G. Summarized results of evaluation including tests, treatments, and measurements.
H. Rehabilitation/Treatment goals
I. Treatment plan
J. Rehabilitation potential
K. Expected frequency, duration of treatment
L. Work restriction

5. Interim Reports/Daily Care
   A. Date of service
   B. Completed after each treatment
   C. Modalities rendered
   D. Any physical complaints
   E. Any treatment precautions
   F. Changes in treatment or work restrictions and goals stated and why
   G. Equipment or supplies administered
   H. Recording of progress, if any, toward treatment goals
   I. Therapist's recommendations and anticipated number of treatments for recovery
   J. Therapist's signature
6. Monthly Summary Report
   A. Date of the progress note
   B. Specific and objective evaluation of patient's progress and response to treatment during the period covered by the note (changes in range of motion, pain level, coordination, strength, ambulatory ability, work capacity)
   C. Changes in medical/mental status, which must be documented in clear, concise, objective statements
   D. Changes in treatment, and work restrictions plan with rationale for the change(s)
   E. Home program and patient education
   F. Therapist's signature
7. Discharge Summary
   A. Appropriate date
   B. Appropriate records of progress made during treatment, functional level at time of discharge, and work status
   C. Recommendations including follow-up contacts, home program, and fitness/prevention (correct postures and exercises).

D. Discharge diagnosis, prognosis
E. Any equipment or supplies required
F. Therapist's signature
G. Filed within 2 working days

Comments: ____

____

5

How the Occupational Health and Rehabilitation Pie Is Divided

REHABILITATION ON-SITE STAFFING POLICY

Growth strategies include the importance of properly delivering on-site industrial rehabilitation services in accordance with the requirements of professional organizations [the American Physical Therapy Association (APTA) and the American Occupational Therapy Association (AOTA)], certifying organizations, licensing bodies, and legal departments. Professional outcome, activities, duties, and responsibilities pursuant to an agreement should be continuously monitored. For example, an athletic trainer is not a physical therapist. The worker receiving physical therapy treatment must have a physical therapist present at the on-site clinic at all times. A physical therapist should never allow an athletic trainer to misrepresent himself as a therapist. Occupational medicine physicians are aware of the differences between trainers, aids, assistants, and technicians. Good assessment and treatment includes keeping a license to perform a scope of practice. A purchasing representative does not know the academic and legal ramifications of substituting one for another. Unfortunately, cost to the patient/worker drives the decision. Educating non–health care purchasers is an ethical responsibility of the physician and therapist.

Therapists must have experience with the Americans with Disability Act (ADA). Occupational and physical therapists should be experienced in resolving conflicts between the mandates of the ADA and those of other laws and inconsistencies among the ADA, union contract terms, and health-related distinctions in insurance coverage. Therapists have experience with making reasonable accommodations, and companies have set guidelines for what is reasonable.

According to Leigh et al. (1997), ''approximately 6,500 job related deaths from injury, 13.2 million non-fatal injuries, 60,300 deaths from disease, and 862,200 illnesses are estimated to occur annually in the civilian American workforce. The total direct ($65 billion) plus indirect ($106 billion) costs were estimated to be $171 billion. Injuries cost $145 billion and illnesses $26 billion. These estimates are likely to be low, because they ignore costs associated with pain and suffering as well as those of within-home care provided by family members and because the numbers of occupational injuries and illnesses are likely to be undercounted.''

Costs to employers in lost productivity, a worker's salary and temporary replacement, and adjudication of disputed claims has caused the demand for occupational health and industrial rehabilitation to increase at the national level. Excessive overexertion injuries (sprain and strain) are second only to flu and colds as a cause of absenteeism (Herrin, 1992). The impact of musculoskeletal disease in the United States is indicated by the frequency of visits to physicians and the frequency of surgical procedures. Low back pain is one of the most common workers' conditions.

The ADA prohibits discrimination in employment practices due to disability. The key obligation of employers is to make facilities readily accessible to and usable by workers with disabilities. Registered occupational therapists acquire or modify equipment and devices for workers and employers. Rehabilitation professionals are asked to perform an increasing number of on-site analyses to identify the risk factors that exacerbate employee injuries. The growth between 1997 and 1998 of on-site occupational health and industrial rehabilitation services provided by the author was 168%. Between 1989 and 1998 the author's on-site occupational health and industrial rehabilitation program grew 2587.7% (measured in net patient revenue). The market is untapped at this time for on-site occupational health and industrial rehabilitation. Company employees are working to age 67. While working, the aged employee will potentially be diagnosed with a rehabilitation disability since conditions of the aged such as arthritis will continue. Insurance companies are looking toward hospital-based programs to manage utilization and for timely access for workers' compensation claims. Labor organizations are writing health and safety issues into union collective bargaining contracts. Occupational health and industrial rehabilitation continues to be driven by the escalation in costs due to poorly managed claims, direct wage replacement, and the need to address the complicated and changing workers' compensation regulations. In addition, employers desire to meet the legal and moral obligations of caring for employees who are injured at the workplace. The 24-hour emergency services on site or at multiple locations and referral to specialists (e.g., plastic surgeons, orthopedists, radiologists) are part of the pie. The on-site occupational health and industrial rehabilitation model has program structure, common processes, and the resources to expand to meet the needs of the purchasing community. Occupational health is also paid generously as compared to other private and governmental programs. Dedicated occupational health experts with a quality reputation and responsiveness to the needs of the company are critical for success.

Customizing programs based on industry characteristics, injury types, previous workers' compensation experience (OSHA log documentation), and an organized structure developed around a core case management team allows the gradual development of protocols and relationships. Mechanisms should be developed for results tracking and compilation that will serve as a link for existing and new business. Opportunities for expansion and to provide product-line managers with program progress evaluation must be identified.

A clinical administrator with hands-on clinical experience and appropriate professional board certification should administratively oversee the scheduling of all company employees. (See Appendix 1 for examples of industrial rehabilitation job descriptions.) The on-site provider should provide licensed, board-certified, and registered practitioners to provide all services hereunder. Therapists shall perform work timely, diligently, and to the reasonable satisfaction of the company in an efficient and economical manner consistent with the best interests of company employees.

The on-site rehabilitation center should be staffed by rehabilitation professionals from outside the corporation and its union partners. At least one licensed physical therapist or occupational therapist with experience in the treatment of industrial injuries should provide direct on-site supervision of the industrial rehabilitation program.

Prevention programs and proper reconditioning should be provided. Programs are available on site to meet the current and future needs of the company. The programs are designed to reduce health costs for the company and their employees without reducing quality of care.

Professional outcome, activities, duties, and responsibilities pursuant to an agreement are continuously monitored. Day-to-day operations of the company on-site therapy and industrial rehabilitation services and responsibilities related to compensation, complaints, evaluations, and attendance are provided by the on-site team. The on-site occupational health and rehabilitation team provides the company with a designated contact person to act as a liaison for the organization. This enhances the provider's ability to serve the organization more effectively.

An experienced administrative director who is responsible and accountable for the management, staffing administration, and operation of the components for which the bid is submitted is the overseer of the on-site program. This individual or a designee should be readily available on a 24-hour basis. The administrative director should be a licensed rehabilitation services provider (PT/OT). All staff will follow the laws of the American Medical Association, Joint Commission on Accreditation of Healthcare Organizations, and state and federal rules for on-site industrial rehabilitation services and have appropriate licensure and/or credentials. The physical therapists and occupational therapists will be skilled in musculoskeletal or soft tissue injury. A certified athletic trainer may assist the therapist with treatment and/or be responsible for the prevention wellness/fitness center located on site at the company. Appendix 1 contains job descriptions for industrial rehabilitation physical therapists, occupational therapists, and athletic trainers.

The administrator will report to the company-designated contract representative on administrative matters and will interface regularly with this individual. The administrator will be responsible for the timely delivery of required management reports. Job descriptions for all professional and clerical staff shall be provided to the company-designated contract representative.

Human Resources/Staffing Benefits

The on-site provider should be able to attract and retain high-quality physical and occupational therapists. Student teaching affiliations for physical and occupational therapy education could be established with nearby universities.

The utilization of services should be controlled and monitored by the on-site team. For example, with the diagnostic related group (DRG) mechanisms, a set number of procedures or outpatient visits are allowed based on a diagnostic code issued for a work-related injury. Outcome measurement drives the termination or continuation of services. The case management office approves or disapproves all treatments based on the initial evaluation and set standards of care. Occupational medicine diagnostic codes with associated costs can be compared against the national averages. Alternatives in pricing and fee arrangements can be explored jointly by the company and on-site provider in order to control costs. Customizing fee structures to share in the risk of meeting medical management, quality of care, and cost containment with company partners or third-party administrators is one way to manage care.

Supervision

Communication between occupational medicine physicians and therapists can resolve questions regarding injured or ill workers' capabilities. Supervision is defined as the authoritative procedural guidance by qualified personnel for the accomplishment of a function or activity within a person's sphere of confidence, with initial direction and periodic inspection of the actual act of accomplishing the function or activity. A qualified person must be on the premises of the person performing the function or activity [qualifications as specified under the Medicare Regulations No. 5, sub-part Q part 4051716(d)].

Administrative Directive

Physical therapy and occupational therapy evaluations on initial visits to all patients will be performed as prescribed. Referrals are accepted and scheduled appropriately for available services. Treatments are performed by qualified physical and occupational therapists along with athletic trainers, physical therapy assistants, and rehabilitation technicians. Routine exercises and other related services used to increase function and prevent injury are performed by physical therapy assistants, certified athletic trainers, and technicians and are carried out while the physical and occupational therapists are on the premises. Patients are evaluated and progress recorded by the physical and occupational therapists and reported to the on-site occupational medicine physician. Adequate progress notes are maintained on all patients and contain statements that reflect the degree of attainment of goals and objectives and the need to change or reassess goals or objectives as appropriate to each patient's progress.

Equipment and Supplies

The company should be informed that all medical and office supplies and equipment specific to providing on-site physical rehabilitation services will be supplied and maintained by the on-site provider team. The facility necessary for performance of these requirements will be provided by the company. The on-site occupational health and industrial rehabilitation team will monitor and maintain inventory management responsibilities associated with said equipment and supplies (e.g., laundry).

Mechanically accurate and properly functioning equipment is essential to patient care. Adequately maintained equipment ensures that treatment will be administered within prescribed limits (e.g., temperature, intensity, or speed). Properly functioning equipment facilitates efficient service delivery and results in maximum use of the therapist's time along with minimum waste of the patient's time. Although equipment failures and problems do occur, a preventive maintenance program can reduce or perhaps eliminate difficulties. It is usually policy to require that electrically powered patient care equipment be regularly inspected, adjusted, or repaired as needed. This may include general maintenance by personnel employed by the on-site occupational health and industrial rehabilitation provider or outside inspection, adjustments, or repairs by specialized service providers as may be appropriate. Manufacturers' recommendations shall be followed. A guide to equipment inspection shall be kept on site for reference. Inspection and maintenance shall be documented.

Routine inspection and maintenance done by on-site personnel (or their contractor) shall be recorded on an equipment inspection and maintenance form dated and signed by the inspecting person. Equipment adjustments, inspections, maintenance, or service repair shall be documented by retention of statements if the services have been performed by a specialized service provider. Inspection, maintenance, and repair records shall be kept on file on the premises. It shall be the responsibility of the administrative director or the supervisor of physical and occupational therapy of the company industrial rehabilitation facility to implement these policies and procedures in consultation with the administrator.

All electrical equipment is to be inspected for performance and safety on a semi-annual basis and is so indicated with a stamp of compliance. If any major repair or replacement is required, the administrative director will be notified for immediate action. All equipment that is found to be defective or malfunctioning will be removed from the treatment area until repaired.

Many on-site service modalities can be supplied by the on-site provider. The equipment and techniques that are usually quoted in company requests for proposals for on-site industrial rehabilitation includes the following:

Ultrasound
Whirlpool
Paraffin bath
Lumbar MedX treatment
Customized hand splints
Myofascial release and
Triggerpoint release
Function capacity evaluation
Work conditioning
Job-site analysis
Back school
Electrical muscle stimulation
Iontophoresis
Phonophoresis
McKenzie technique
Diathermy
Jobst pump
Massage
Hot/Cold therapies
Traction: cervical/pelvic
Ergonomics
Prevention and proper posturing

Additional services may be required at the company as deemed appropriate based on company injury/illness demographics.

In addition to modalities, the following treatment and evaluative devices may be of value:

1. Whirlpools/Chairs
2. Hydroculator—hot/cold
3. Ice machine
4. Treatment tables—wood
5. Weights, dumbbells with rack
6. Measurement tools: goniometer, dynamometer, pinch gauge, volometer
7. Treatment mats
8. Fitness equipment/Nautilus 15 Station or Cybex and MedX equipment
9. Step aerobic machine

10. Treadmill
11. Bike: Airdyne; Lifecycle
12. Carts
13. Traction table/equipment
14. Electro muscle stimulator with cart
15. Isokinetic apparatus
16. Work conditioning equipment
17. Ultrasound
18. Baps board systems
19. Physio balls—1 large, 1 small
20. Hand-splinting equipment (heat gun, hydrocollator, finger and wrist prefabrication and fabrication, e.g., polyflex/orthoplast materials)

The following suggested list of equipment and supplies may be supplied by either the on-site provider or the local plant:

1. Health information system computer and software capabilities
2. Telephone and message recorder
3. Restroom facilities
4. Fax machine
5. TV, VCR
6. Copy machine
7. Refrigerator/Microwave
8. Bulletin board
9. Coat rack, chairs
10. Desk, file cabinet, desk chair, computer table, miscellaneous office items (with locks)
11. Unisex changing room equipment (e.g., lockers)
12. Laundry service
13. Therapeutic supplies
14. Others, as deemed appropriate

The on-site provider team implements a performance review program, based on staff job descriptions. A formal quality assurance program is planned for implementation at the initiation of the program, including medical treatment protocols, anticipated outcomes, and utilization review.

The on-site provider team should supply the company (i.e., plant) with a policy and procedures manual that defines acceptable professional practices. This will include physical and occupational therapy as well as athletic trainer practice and directives for treatment service assigned and dated by the occupational medicine physicians assigned to the company. Periodic review and an update of the manual is completed annually. Policies and procedures for therapy guidelines are shown in Appendix 2, which includes policies and procedures for protection of clinical records, referral policy, clinical record documentation, a sample of infection control, and a guideline for clinical procedures.

The maintenance schedule for physical therapy equipment is important to ensure that it meets standards of operation. Table 1 illustrates the rehabilitation equipment that should be serviced and the frequency of implementing the process. The guidelines for calibration state that all electrical equipment for patient treatment be inspected by a licensed operator for accurate calibration and ground (shock hazard) on a semi-annual basis (on-site personnel will continuously check for electric cord fraying and visual part safety and report to the administrator or supervising therapist). The licensed calibrator will provide a written report, and any correction needed will be followed up at the discretion of the physical and occupational therapist and the administrator.

TABLE 1 Maintenance Schedule for Rehabilitation Equipment

Unit	Maintenance	Frequency of main service
Large hydrocollator	Ground/Calibrated	Semi-annually
	Temperature	Each use
	Cleaning	Each use
Whirlpools	Ground/Calibrated	Semi-annually
	Cleaning	Each use
	Temperature	Each use
Paraffin	Ground/Calibrated	Semi-annually
	Cleaning	Each use
	Temperature	Each use
Ultrasound	Ground/Calibrated	Semi-annually
	Maintenance clean	Weekly
	Sound head clean	Each use
Electric stim.	Ground/Calibrated	Semi-annually
	Maintenance clean	Weekly
	Pad/Sponge Clean	Each use
Cold quartz	Ground	Semi-annually
	Maintenance clean	Per use
Jobst pump	Ground	Semi-annually
	Maintenance clean	Each use
Traction tables	Ground/Calibrated	Semi-annually
	Maintenance clean	Weekly
Tens units	Each unit use/Supplier-consignment	

PHYSICAL ENVIRONMENT

The on-site provider team recognizes the importance of a functional, safe, and healthful environment for patients, personnel, and the public alike. Written policies and procedures shall clarify the standards, which shall be met and monitored. Such policies and procedures shall include, but not necessarily be limited to, those that relate to fire, building, equipment, and grounds. The administrator is responsible for monitoring the implementation of policies and procedures as adopted by the executive board and approved by the board of trustees of the company contracting services, if necessary. The provision of necessary office services, such as copying, transcription, data processing, data handling, secretarial and clerical services, etc., is the responsibility of the on-site provider unless offered by the company. The company (i.e., the plant) should provide customary "landlord" services for the on-site occupational health and industrial rehabilitation facility, including utilities, telephone hook-up service, basic housekeeping, general maintenance, and security services. The on-site provider understands that the replacement or acquisition of additional company-furnished equipment for plant site use is negotiable.

The provision and use of the on-site provider–owned or - leased equipment which the team wishes to use at the plant site will be subject to company approval. Physical and occupational therapists will make appropriate recommendations for physician specialists, independent third-party medical opinion as outlined in the collective bargaining agreement for employees with occupational and nonoccupational sickness, and accident disability leave if requested. All patient cases are discussed in grand round case management meetings.

The transitional work program consists of a temporary modified work program process with the objective to provide "temporary" modification of an employee's occupation while on restricted or limited duty status during a course of treatment, therapy, or a work reconditioning program designed to return the employee to the full scope of the employee's occupation. Qualification for temporary modified work programs (TMWP) shall be determined on a case-by-case basis and approved by the on-site occupational medicine physician, department supervisor, and the industrial relations department. Where the company on-site occupational medicine physician concludes that TMWP is medically appropriate, the company physician shall discuss the restrictions or limitations with the industrial relations department and the supervisor of the employee's department. The injured/ill worker's supervisor evaluates the department's ability to return the employee to a modified job in his or her department. This includes an evaluation of the effect of restrictions or limitations on (1) the employee's ability to perform the job satisfactorily, (2) the productivity of the operation, (3) the ability to shift work or modify work to accommodate the restrictions, and (4) the probability of further injury.

The duration of the TMWP shall be limited to a period of 4 weeks, which can be extended with approval of the attending physician, company physician, department supervisor, and the industrial relations department. An extension of up to 4 additional weeks *must* be based on a reasonable assumption that the extended period will result in return to the full scope of the employee's occupation. On-site treatment, therapy, or physician appointments shall be scheduled during off-shifts, i.e., prior to or after the injured/ill employee's work shift. In the event the employee does not qualify for TMWP, nothing in the process will preclude the employee or the employer from exercising their rights under the labor agreement or any applicable state or federal laws. TMWP shall be available under the same parameters, on a voluntary basis, for nonoccupational disabilities.

The following process describes both union and management joint efforts to implement a modified work program. In the latter months of 1999, a special meeting was conducted at one of the steel industry headquarters to discuss workers' compensation problems that were being experienced at three steel facilities. Present at the meeting were the human resources managers and the safety supervisors from the plants as well as corporate human resources personnel. Much of the meeting was spent discussing formulating a modified work program that was adaptable to the facilities. Individuals were appointed to draft a usable modified work program. A temporary modified work process was developed from the meeting.

In order to adapt a process to the plant, modifications were made. Qualifications were approved by the safety supervisor, the area supervisor, and/or operations superintendent based on information provided by the on-site occupational medicine physician. Intentions to implement the TMWP on a case-by-case basis was reviewed with the union grievance chairperson. A functional job requirement list (physical demands of job) was generated for the jobs in question and then matched with employee's restrictions and job preference. As a courtesy and to lay out the parameters of the case, the on-site occupational medicine physician was required to generate a physical capabilities checklist of the injured employee to be used by the area supervisor, safety supervisor, and general supervisor to determine if meaningful work could be provided to the injured person. In addition, the physician was required to give an opinion as to whether or not the employee would be capable of full return to work within one month, with the option of an additional month of modified work if necessary. The superintendent evaluated the department's ability to return the employee to meaningful work.

The determination for the employee to return to meaningful work in the preceding example was based on (1) the list of the employee's restrictions, (2) the functional requirements sheets for each job in the department, and (3) the area supervisor's account of the physical requirements of each job in his or her department. This process decreased the above company's indemnity payments by 54%.

TMWP is limited to 4 weeks, which can be extended by written evaluations from the on-site occupational medicine physician based on reasonable assurance that the employee is capable of a full return at that time. In the event that no modified work is available in the preferred original job, the case is reviewed with the operations supervisor to determine if work is available within the general labor pool. If so, the employee's modified work program is administered by the general supervisor. Treatment, therapy, or physician appointments are scheduled during nonwork time. Should a schedule conflict arise, the department supervisor may arrange to have work schedules changed. Temporary modified work will be available under the same rules and guidelines on a voluntary basis for sickness and accident cases as well. Loss of wages due to variances in pay scales between job worked on modified work and preference job available will be reimbursed through the third party administrator according to the state workers' compensation statute. The area supervisor will submit a weekly ''partial disability'' form (Fig. 1) to the safety supervisor, who will forward the form to the payroll department. Payroll will tabulate the earnings for the week and the report will be forwarded to the insurance carrier.

Usually the safety supervisor chairs a review of the TMWP semi-annually. The plant manager, operations superintendent, manager of industrial relations, and union grievance chairperson are invited to the review. Each TMWP participant's case is reviewed. A yearly update is sent to the employees via an interdepartmental memorandum, as illustrated in Appendix 3.

Work reconditioning addresses the major issues in work conditioning that have been brought to the attention by organizations, surveyors, and founders. The program description for work reconditioning uses the term ''interdisciplinary,'' which is defined as being characterized by a variety of disciplines that participate in the assessment, planning, and/or implementation of a patient/employee program. There must be close interaction and integration among the disciplines so that all members of the team interact to achieve team goals (e.g., physical therapist, occupational therapist, occupational medicine physician, occupational health nurse, psychology, vocational rehabilitation). The program description also outlines the areas that must be addressed, including functional, physical, behavioral, and vocational areas.

Services by all disciplines on the on-site team should be provided. They may be provided by full-time or part-time employees or through contract or other formal arrangement. If all services are not provided or all disciplines are not actively involved, the company may be providing hourly workers as job site coordinators and placement workers as substitutes. Job placement, ergonomics, plant health and safety, and human resources are kept up-to-date by the occupational therapist providing work reconditioning. Again, it is imperative to have appropriate licensed and board-certified clinicians performing work conditioning program oversight on a daily basis on site.

Employee Name:______________________________

Check Number:_________________________

Date of Injury:___________________ Nature of Injury:____________________

Date employee returned to temporary modified job:_______________________

Job Classification (Preference job available): __________________________

Normal Rate of Pay: __________________________

Job Classification (Temporarily Assigned): __________________________

Present Rate of Pay: __________________________

Date employee resumed regular job or the same rate of pay prior to the injury:

________________________ __________________________

General Foreman/Date

Please return to the Safety Office for forwarding to the Insurance Carrier:

FOR PAYROLL USE:

EARNINGS:

Week Ending:_______________ Gross Earnings: _____________________

____

Authorized Signature/Date

Figure 1 Partial disability form.

Services include screenings, assessments, evaluations, program planning, treatment, training, and other specific techniques. The psychological and vocational guidelines for return to work are considered important for the well-being of the patient/employee recovery. In addition, there should be documentation that indicates that each person served has been screened/evaluated in order to determine the need for all of the services. If a screening tool is used, it should be developed by the discipline that will be interpreting its results. That discipline should also be providing the training to staff on how to administer and interpret the screening if they themselves will not be administering it. Knowledge of medical conditions and contraindications are a must to allow for patient recovery.

Group work with no screening or assessment of the needs of the patient served does not meet the company's needs. Group work that comes from a need that has been determined in the screening/assessment process is appropriate. This should be evident through documentation and discussion.

On-site industrial rehabilitation guidelines include the following points:

There is no particular square footage requirement for work reconditioning. It is suggested that there be a designated area for the program.

No company approves or endorses any one type of work reconditioning equipment.

The work reconditioning program may be one component of an on-site occupational health and industrial rehabilitation facility. Organizations are encouraged to clarify the distinctions among their programs.

Work reconditioning is a treatment program of general conditioning and/or work simulation designed to raise the client's physical level to the highest possible work range.

The number of people served in the work hardening or work reconditioning programs has been questioned when it appears to be extremely limited. There does not need to be a minimum number of persons served or required to qualify as a work reconditioning program. If the numbers are extremely limited, however, there may be difficulty in the organization's demonstrating a valuable service to the company.

Work reconditioning has prevented reinjury and prepared the patient/employee for placement back in his or her original job classification.

Work reconditioning is highly structured and goal oriented. It is a transition between initial injury and return to work. Work reconditioning uses real or simulated work activities in conjunction with physical and occupational therapy. Registered, board-certified occupational therapists structure therapeutic activities available 5 days per week for up to 6–8 hours a day. The referral criteria for an injured/ill worker consists of stringent criteria. The patient/worker must benefit from the program design. The injury/illness should not interfere with carrying out job-related duties in the workplace. The worker should not have medical, psychological, or other conditions that would prohibit him from participation in the program. The referral process should include anticipated goals that would return the person to work and time frames within which to work. A personal interview of the employee's work history, current functional status, restrictions, educational background, and recent medical status will be conducted by the registered occupational therapist. The evaluation process will assess the worker's behavioral, attitudinal status, motivation, cardiovascular status, cognitive status, functional work capacity, musculoskeletal status, work history, and current job restrictions. Clinical assessment evaluation reports will document the job activities completed, measured results achieved, and the duration of reconditioning program. The interdisciplinary team will consist of a registered occupational therapist, physical therapist, job placement coordinator, and plant/company occupational medicine physician with demonstrated training and experience in work reconditioning, functional tolerance screening, ergonomics, and on-site job assessment. Services will be coordinated and grand rounds/case management will be held at least every other week. Additional services available may include strengthening, flexibility, stretching, weight loss, nutrition, and back or hand classes. Other services such as industrial engineering, safety, ergonomics, the fitting of orthotics/prosthetics, nursing, and employee assistance programs (EAP) are available on site. The employer and insurer are involved in goal setting and reinforcing work reconditioning. These efforts must be documented at grand rounds/case management. The program should increase strength and endurance in relation to work duties and safe job performance skills. Specific job requirements are reviewed and the education of the employer and union is ongoing. Productivity, safety in the workplace, and workers' musculoskeletal requirements and capabilities are documented by the registered occupational therapist.

The following examples illustrate an industrial rehabilitation on-site job assessment, job coaching, functional capacity evaluation, and job task analysis and evaluation used to familiarize and assist on-site industrial therapists in formatting reports. Samples are merely used for application purposes, not for content.

EXAMPLE 1: OCCUPATIONAL THERAPY—ON-SITE JOB ASSESSMENT AND JOB SITE COACHING

Data Base

Patient Name: Sue M. Medical Record Number: 37568909

Date: ______

Diagnosis: Cervical Neuropathy C5-C6 & Cervical Compression (left)

Onset Date: ______ Ins: Workers' Compensation

Physicians: Dr. Kim and Dr. Lee

Therapist: Joanne Stone, OTR Telephone Number: (313) 874-0000

Benefits Representative: Ms. Cathy C.

Location: Banking Industry, 593 Street, Detroit, MI 48202

Supervisor: Ms. Janene W. Job Title: Payroll Teller

Date of Birth: ______

Medications: ______ Past Medical History: ______

Job Summary

Receives, verifies, processes, straps, and disburses currency. Mrs. M's current duties primarily entail processing, per restrictions. Thus, her usual and customary job duties involve rotation on a weekly basis throughout three different operations. These job operations are as follows:

1. Payroll teller
2. Preparation teller
3. Bending, bundling checks, and currency use of the jogger machine and bending to turn brake on/off under the machine

Mrs. M has a stool in her cage but indicated that she was unable to sit while working because of the pace of the processing machine.

Production

The job rotation consists of an 8-hour shift of maintaining the continuous pace to prepare, process, strap, wrap, bag, handle, and disburse money to meet critical job demands. According to the supervisor, sums processed vary from $15 to $100 million, depending on the day of the week.

Physical demands

At the time of the evaluation, patient was able to perform light duty work as processor despite complaint of pain/discomfort. A major concern noted was patient maintaining posture of head/neck in flexion for a long period of time while verifying, strapping, and wrapping bills.

Other physical demands include:

1. Constant standing and walking from counters to machine to maintain production
2. Prolong head/neck flexion with handling money
3. Constant reaching and grasping of currency to process, band, and wrap currency
4. Occasional to frequent bending to lift varying amounts of cash
5. Occasional to frequent handling of cash weighing 20–35 pounds per bag
6. Occasional pushing/pulling of cart to transport cash to various areas; maximum weight on the cart was 735 pounds

The jobs within the three areas ranged from light to medium work levels, lifting 20–35 pounds frequently during each shift in addition to lifting approximately 25 pounds from floor to overhead (fast-cash area). Also, employees are required to push in excess of 750 pounds varying distances.

Recommendations

1. Reorganize job tasks to allow for frequent change of positions. This would prevent prolonged head/neck flexion or repetitive upper extremity use, eliminating muscle fatigue and pain.
2. Arrange shelves so that the heavier loads are between shoulder and waist height. Lighter, less frequently used items should be placed on higher shelves and items rarely used on lower shelving when possible.
3. Alternate sit/stand posture to reduce back and lower extremity stress.
4. Keep money bags close to body to eliminate excessive reaching.
5. Provide foot stool, placing one foot up on stool to decrease back and leg fatigue.
6. Provide two- to three-step ladder (rolling) with safety rail for increased balance rather than step stool. This is to access overhead shelves in the fast-cash area, decreasing excessive reaching above head.
7. Redesign vaults with installation of two-steel shelves on roller, with two drawers on each shelf for easy use. Space lost is less than 1 inch inside of vault. This will eliminate excessive reaching and bending to access the money bags.

8. Employees may pace using the lumbar supporter issued by the department for low back support with bending and lifting tasks.

The above recommendations are potential solutions for reducing the risk of injury or reinjury while enhancing employee productivity. The different job locations, limited cage space, and security needs with regard to storage units and vaults have been considered in the overall analysis.

Assessment

Mrs. M appears to understand the importance of utilizing the correct posture for lifting, reaching, material handling, and bending in addition to the need to pace herself while performing duties to decrease pain and prevent reinjury. Her endurance was good for the evaluation.

Plan

The patient indicated that she is to continue in physical therapy, which at present has been arranged during her work schedule. She is currently capable of performing restricted job duties. She will follow up with physician's visit and the results of the electromyelogram (EMG).

Thank you for this referral.
Respectfully submitted,

Registered Occupational Therapist:______________

EXAMPLE 2: INDUSTRIAL REHABILITATION OCCUPATIONAL THERAPY: INITIAL EVALUATION

Data Base

Patient Name: Mrs. J ________ Social Security Number: ______________

Diagnosis: Sciatica

Place of Employment: On-site Industrial Assembly Plant

Referring Physician: ______________________________

Date of Initial Evaluation: ______________

Dear Dr.: ________________________

Mrs. J was initially referred to us on January 30, 1999, however, she was unable to participate in an evaluation prior to today's date because she had been removed from the plant for disciplinary action. On January 30, 1999, Mrs. J was seen in the Medical Department, and the nurse there referred her to the Transitional Work Center Area for an evaluation. Mrs. J was somewhat cooperative and made an attempt at each task requested of her. According to the U.S. Department of Labor Job Classifications format, Mrs. J is currently functioning at the ''light'' work demand, with the ability to lift 20 pounds infrequently and up to 10 pounds frequently. Her job as an assembler performing the gas fill job requires lifts of 5 pounds or less at waist level. This is classified as ''light work.'' Please review the following summary for further details.

Performance Measures

Mrs. J's upper extremity range of motion and strength were found to be within normal limits during manual muscle testing.

Mrs. J has a right-handed grasp of 76.66 pounds and a left-handed grasp of 66.66 pounds. These grip strength measurements place her above the 90th percentile when compared to other females of her age. Mrs. J had a lateral pinch for the right hand of 17 pounds, and the left hand is at 14.66 pounds. These lateral pinch measurements placed above the 90th percentile for the right hand and in the 75th percentile for the left hand. Mrs. J's 3-jaw-chuck pinch for the right hand measured 14.66 pounds, and the left hand measured 14 pounds. These 3-jaw-chuck pinch measurements placed her bilaterally in the 75th percentile.

The patient's manual dexterity skills were tested using the Minnesota Rate of Manipulation, the Purdue Pegboard, and Valpar Eight. The patient scored poorly on the Purdue Pegboard, placing only 11 pins with the right hand in 30 seconds and 13 pins with the left hand in 30 seconds. She scored below the 1st percentile for right- and left-handed placing on the Minnesota Rate of Manipulation. It took her 150 seconds to complete the placing tasks for the right hand and 162 seconds to complete the placing tasks for the left hand. Interestingly, on Valpar Eight, which is a three-piece assembly task, the patient was able to complete 178 three-piece assemblies in a 20-minute time span. This score placed her in the 60th percentile for a three-piece assembly task. The patient does display a bilateral hand tremor.

The patient's spinal and pelvic girdle flexibility was measured. She is able to achieve 105 degrees of forward flexion at the waist. She was able to achieve 25 degrees of hyperextension at the waist. Mrs. J was able to achieve 15 degrees of lateral flexion to the right and only 12 degrees of lateral flexion to the left. She was able to perform a straight leg raise from a supine position with the right leg of 65 degrees, and the left leg was able to achieve 75 degrees. She was able to achieve 115 degrees of quadriceps flexion bilaterally. She was able to attain 90 degrees of right hip flexion and 100 degrees of left hip flexion while in a supine position. The patient was able to attain 45 degrees of low back hyperextension in a prone position.

During maximal effort testing, the patient was able to achieve the following lifts: floor to knuckle 20 pounds, 12 inches to knuckle in lordosis 25 pounds, knuckle to shoulder 21 pounds, and shoulder to overhead 15 pounds. Additionally, the patient was able to carry 20 pounds a distance of 28 feet. She also carried 25 pounds a distance of 28 feet, but reported an increase in low back discomfort with the addition of 5 pounds. The patient was able to push and pull a sled requiring 30 pounds of push/pull pressure a distance of 15 feet in each direction. Additionally, the patient was able to perform repeated bending at the waist to within 12 inches of the floor 35 times in a 3-minute time span. She was able to squat like a baseball catcher for 30 seconds before stating that she was afraid that she was going to lose her balance.

The patient states that she is able to walk up and down a flight of stairs in her own home. She was observed walking to and from the cafeteria with an approximate distance of 1/4 mile. She has a normal heel strike gait. She also performed a variety of tasks while seated, and she was able to sit comfortable for a 30-minute time span.

Past Medical History

Mrs. J is a 46-year-old female who is 5 foot 5 inches tall and weighs 230 pounds. She states that she has had episodes of back pain over the last 4½ years with no one specific precipitating incident. The patient states that she has had a CAT scan but is uncertain as to the results. She also states that she has had an electromyogram (EMG), which indicated that she has a pinched nerve.

Mrs. J also states that she has had numerous x-rays, but she is uncertain as to the results of those x-rays. The patient states that she has received injections into her low back hip area at the pain clinic in the hospital. She is returning to her physician on February 14, 1999.

The patient states that she has participated in physical therapy at the assembly plant. She was unable to identify the home program given to her by the physical therapist. She states she has been involved in a car accident in the past.

She has had past surgeries including gallbladder and a partial hysterectomy. Mrs. J states that she is presently taking Feldene, Elavil, and Lorcet. She states that she has been out of the assembly plant due to a nervous condition, but did not elaborate on this condition.

Mrs. J states that she experiences aching across her low back and sharp pain in her hip area. She states that she also experiences aching in her right frontal thigh area. Mrs. J states that her pain right now is a 4, at its worst a 9, and at its best a 2 (this is based upon a 1 to 10 pain scale, where 1 is minimal pain and 10 is pain severe enough to necessitate a trip to the emergency room). She states that her pain symptoms begin or increase when she starts working. She cannot "bend, lift, or at times walk or stand in one spot without increasing her pain symptoms."

Summary/Comments

Numerous inconsistencies were revealed during Mrs. J's evaluation. She states that she can only sit for 30 minutes, yet she can drive a vehicle for 4½ hours, an activity that requires sitting. Mrs. J stated that she cannot reach to pick an item off the floor, yet she demonstrated the ability to do so during today's evaluation. She also stated that she could not crouch like a baseball catcher, yet she performed this activity for 30 seconds before complaining that she might lose her balance as opposed to limiting the activity due to any pain. Mrs. J also had an abnormal lift pull index of 1.75, with normal being between 2.0 and 3.0. Mrs. J reports an increase in pain with pseudo-trunk rotation. She also complains of superficial skin tenderness in the area of her upper hip.

Although the patient reports that she has been experiencing back pain, she has not been participating in her home exercise program provided to her by the physical therapist. When asked which exercises she was to be performing at home, she stated "Oh, if only I can remember them." Unfortunately, this patient was unable to recall the exercises assigned to her by the physical therapist.

Recommendations

The patient states that she is going to see her own neurologist on February 20, 1999. Hopefully, this visit to her neurologist will shed more light on her subjective complaints. Mrs. J is obese and would benefit from a structured weight reduction program if she is motivated to undertake such a program at this time. Mrs. J by her report has not been completing the home program provided by the physical therapist.

At the present time, a month has nearly elapsed since the referral of this patient to the transitional work center area. If the referring physician would like continued therapy in the form of reconditioning, this can be initiated once she is reevaluated by the medical department at the assembly plant. As a part of today's initial evaluation, the patient did view an audiovisual presentation for education on the use of proper body mechanics.

We would like to thank you for the referral of this patient to the transitional work center for an initial evaluation. If we can be of any further assistance at a later date, please feel free to contact us.

Registered Occupational Therapist:______________

EXAMPLE 3: INDUSTRIAL REHABILITATION JOB TASK ANALYSIS

Company: Manufacturing Industry

Patient Name: ___________

Job Title: Inspectors: 1) polisher, 2) tester, 3) #11 line

Date: ___________

Date of Injury: Status Post 6 Weeks

Diagnosis: Fracture of Ankle with Injured Peroneal Tendon

Contact Person: Mr. John Smith

Telephone No.: ___________

Registered Occupational Therapist: ___________, MA, OTR

Past Medical History: None

Medications: None

Schedule

Shift(s): Day/Afternoon/Night (rotation every 2 weeks)

Hours/Week: 8 to 16 hours (mandatory overtime with no relief person)

Key Job Tasks

Polisher Inspector—Is required to know operation of measuring equipment. Inspect coil for defects, operate computer, know manufacturing standards, spill prevention control, and countermeasures; crane safety and mobile equipment safety; also pinch point of unit.

Tester Inspector—Is required to know sample collection and preparation of steel, operate computer, also various machines—Sin Tech, Tensile, Baldwin, Olson, Bender and Buffing; also required is the testing of steel for strain and height; performs Rockwell hardness and verification tests and the Bent and Rbar Tests.

#11 Line Inspector—Is required to know operation of on-line computer system for monitoring steel for defects; also, know operation of ''OS-18, OS-1 rehydraulic and water systems,'' steering furnace, fumes, and motor blower steering; inspector on #11 line must know pinch points of unit.

Work Pace

Self-Paced: Polisher and Tester
Incentive Based: Yes
Machine Paced: Inspector #11 Line
Quota System: Daily average of 5–6 coils of light-gauge steel, 10 coils heavy-gauge steel. Weight of coil range = 15,000–30,000 pounds per coil
Other: #11 Line constantly running (inspector states she seldom leaves station).
Equipment, Machines, Tools, Vehicles: Buffer, Olson, Rockwell, SinTech, Tensile, Baldwin, OS-1, S-18, pendent crane, grinder shearer, and hand truck.

Environmental Exposures

Inside: 100% Outside: Temp. Range: 70–80°F
Comments: Noted steam constantly escaping from wash tank in the polishing area, obstructing visibility in this location.

Chemicals: Contact

1. Consumption comments: fumes from chemicals especially in polishing and #11 line areas. There are three acid tanks situated approximately 20 feet from inspector's station at #11 line.
2. Inhalation comments: Coolant and acids—hydrofluoric, nitric and sulfuric acid fumes noted in polishing area and #11 line (somewhat slight to moderate).
3. Confined space comments: Polishing and #11 Line areas were crowded with two or more people.
4. High elevations comments: In the #11 Line area, the inspectors are required to climb eight steps to the work area. Polisher inspector is required to climb up/down three flights of stairs/platform to the grinder on days.
5. Iodizing radiation/noniodizing radiation comments: Iodizing radiation from machine (x-ray) in tester's area.
6. Moving objects comments: Overhead cranes approximately 20 feet high, moving frequently to constantly throughout inspector's area. Additionally, forklift, trucks, and coil line constantly moving.
7. Noise comments: Noise level extremely high due to equipment and machinery, which has the potential to cause distraction to workers.
8. Safety equipment comments: Hardhat, safety glasses, steel-toe shoes, and ear plugs available.

9. Slippery surface comments: Concrete and metal grate floors slippery from oil, water, and coil bans on the floor throughout the plant.
10. Special clothing comments: Fire retardant gloves, jacket, and pants are worn.
11. Vibration comments: High level of oscillation due to Hi-Lo, cranes, grinder, air compressor, and other machinery throughout entire plant.
12. Wetness comments: Wash tank spills from coil line, approximately 5–10 times daily, per polisher inspector.

Physical Demands of Lifting/Material Handling

Weight (lb)	Never— 0%	Rare, 1–10%	Occasionally, 11–33%	Frequently 34–66%	Continuously 67+%	Objects	Height
1–10			X			Micrometer, manuals	
11–20							
21–35							
36–50			X			Test samples of steel	Storage unit is 6 feet in height
50–75		X				Trash barrel	3 feet height
76–100		X					
100						Contact rolls and furnace rolls	6 feet
Carrying	Never	Rare	Occasionally	Frequently	Continuous max dist carried 10 feet distance	Objects	
Max, 50			X			Test samples of steel	Approx 6 ft
20			X			Pieces feed steel with bender machine	
Push/pull Max Force	Never	Rare	Occasionally	Frequently	Continuously	Objects	Max dist Moved 10 feet
10–15			X			Table underlined on ropes	Approx 5–7 feet
15–20			X			Hand truck (tester)	Plantwide use

Comments

Patient performs the job duties of inspector tester during most of her shift with lifting and material handling, as noted with steel samples. However, once a week all inspectors are required to handle materials. Polisher inspector must shovel pit (heavy cardboard, coil scraps, etc.), empty trash barrel, replace rolls on paper, operate pendent crane, and change 300–400 pounds of solid rubber contact rolls. The #11 Line inspector must clean out acid tanks and change furnace rolls (furnace temperature ranges 1000–2100°F) and other maintenance duties. It is recommended that additional storage units be considered in tester's area for storing steel samples between waist and shoulder height (4–4.5 ft) rather than overhead.

This will eliminate the risk of injury involved with lifting 50 pounds overhead to store heavy samples of steel.

Postures/ movements	Maximum consecutive min/hrs	Total daily hours	Position change optional	Further description
Sitting		2		Polisher and tester Inspector
Standing		6–8		All inspectors
On feet		8		#11 Line Inspector
Walking		4–6		Tester to pick-up steel plant wide

Postures/ movement	Never	Rarely	Occasionally (hr)	Frequently	Constantly	Further description
Bending			1			Check underneath the script, etc.
Turn/twisting			4–6			Turning and twisting while monitoring coil machine
Kneeling		X				
Squatting		X				
Crawling		X				
Climbing			1–3			Up/down steps and 10-in. step in polisher area. Also, 8 steps #11 line
Reaching out			8			#11 Line inspector, 30–35 hand control buttons
Reaching up			8			#11 Line inspector, 30–35 hand control buttons
Wrist turning			8			Operating machines, control buttons (see Tester, Polisher & #11 Line).
Grasping			8			Coils and bander
Pinching			1			Micrometer, pens/ pencils
Finger manipulation			8			Computer typing

Summary

Based on the U.S. Department of Labor's Dictionary of Occupational Titles–Physical Demand Characteristics of Work, the Polisher, Tester, and #11 Line Inspector's job duties rank at the medium work level—lifting 50 pounds of weight, infrequently. This is noted on ''down days,'' when all inspectors are required to lift and handle 50 pounds or more of materials including trash barrels, contact rolls, and paper and furnace rolls.

Environmental exposures, including fumes from chemical contacts, vibration, and extremely high noise levels, are factors that affect the demands of this job classification.

Thank you for the opportunity to provide the job task analysis.

Registered Occupational Therapist: ______________

EXAMPLE 4: OCCUPATIONAL THERAPY WORK CAPACITY: INITIAL EVALUATION

Date: ___________

Physician: ______________________

Patient: ________________________

Ins. Carrier: Creative risk

Company: ______________________

Diagnosis: Post–lumbar laminectomy

Date of Injury: 9-4-98 Date of Birth: ___________

Onset: Date of injury 9/4/98

Past Medical History: C-section, 1989; tubal litigation, 1990; at age 4 she had a leg fracture; she is unable to identify which leg.

Medications: Patient currently is not taking any medications. She takes vitamin E and calcium supplements daily.

Surgery Date: 12/14/98 at St. Joseph Hospital. She states she had herniated L4 and L5 discs due to a compression injury from lifting a box off a truck trailer. She states she had lifted the box from overhead and positioned it on her left shoulder. She states that the box weighed approximately 50 lbs.

Patient's Job Title: Electrical contractor

Company Name: Guardian Electric

Job Summary: Patient works as an apprentice for Local #58 Contractors. She performs installation of lights, switches, plugs, thermostats, garbage disposals, and all electrical installation for commercial and industrial buildings. Buildings include schools, hospitals, auto industry and electrical contracting.

Description of Tasks: The patient performs a variety of tasks including lifting, climbing, manipulation of small parts, utilizing bilateral upper extremities to handle power drills and tools and other mechanical devices. She also is required to push and pull carts and climb up and down ladders.

Vocational Preparation: Patient is involved in an apprentice journeyman program for perfecting her electrical trade. The educational requirements are 5 years. She will finish her second year in May 1999.

Past Work History: Bartending for 2 years at Jack's Bar in Pontiac, Michigan. Bartending for 5 years at the Dixie Bar in the Pontiac area. She had worked various sales jobs at Kroger, McDonald's, and Hudsons for a combination of 6 months to 5 years of work at each location.

Activities of Daily Living: Patient states that she has a 9-year-old son. She states that he has been a great help to her. She lives in a two-story home with 5–8 stairs to climb to enter the home. She states that she sleeps on a sofa on the first floor and she is not going up and down stairs to reach the second level. She states that she has a bathroom on the first floor. She has her son assist her and uses her son to help her in and out of the bathtub. She also states that her son is helping to perform the grocery shopping by lifting of grocery bags. She states she is independent with dressing, bathing, and hygiene. She wears slip-on shoes. She is unable to tie shoe laces at this time.

Patient Goals: "To be an electrician and to return back to the occupation in this capacity."

Objective: Patient is a 32-year-old female. She is right hand dominant. She drove herself to the clinic today, utilizing her left foot on the gas pedal. She states that when she adds pressure to the right ball of her foot or heel, the pressure sends pain from her foot through her knee. She states "it is a shooting pain." On a scale of 0–10 with zero being no pain and 10 being extreme pain, she states her pain is at level 4. She also complains of pain on the right side of her incision at a level 4.

Body Mechanics: Observed to be fair. She was instructed in how to use good body mechanics as well as maintaining good posture for preventing strain and damage to the joints, muscles, ligaments, and bones. The patient is currently able to sit for approximately 2 hours. She hyperextends her back to stretch and take the pressure off the lumbar region. It is recommended that she receive a lumbar roll to aid in comfort during sitting.

The patient is able to squat, crouch, kneel with assist in going from standing to squatting, crouching and kneeling. The patient utilizes her right upper extremity and braces herself against a wall or a stable table to obtain these positions. While standing, she shifts her weight off her right leg onto her left side. The patient also utilizes a stool to rest her right leg upon when standing to avoid strain and pain over the lumbar region on the right side. Patient showed good body mechanics when bending. She was able to bend at the hips and knees and to use her left leg muscles to lift and keep her body erect. The patient was given instructions on how to use bilateral upper extremities to carry objects and to walk instead of twisting her back in order to face objects ready for lift. She was also instructed to keep objects close to her body with her back in her normal position. The patient previously was never instructed in the correct lifting and carrying techniques to avoid pressure to the lower back area.

The patient was able to stand for 15 minutes during the evaluation. She prefers taking small steps and walking and shifting her weight rather than standing in a stable position. This is recommended to avoid further pain to the lumbar region.

The patient was able to lift a 3½ lb milk crate from waist level to shoulder level, from shoulder to chin level, from chin level to eye level. She was not able to lift overhead at this time. Pain increased to a level 8 upon this motion.

The patient was able to walk 150 ft carrying the 3½ lb milk crate with a 1 lb. free weight inside, for a total of 4½ lb of resist, three repetitions. The patient was not able to walk 150 feet with 5½ lb of resist. At that point the pain level increased from level 4 to a level 8, with sharp shooting pains described in the ball of her foot, proximal to her knee and shooting through the lumbar region surrounding the incision in her back.

The patient is able to lift a 3½ lb milk crate from waist level to thigh level but is unable to place the carton to knee level, ankle level, or floor level.

Unilateral lifts of 4½ lb are tolerated at all levels: floor to ankle, ankle to knee, knee to thigh, thigh to waist, waist to shoulders, shoulders to eye level. The patient is not able to lift/resist from eye level to overhead reach. Pushing and pulling utilizing a cart was performed. Pushing is less strenuous than pulling. Patient had good body mechanics during this push/pull cart activity. The patient was instructed to utilize her legs as much as possible. Pushing the cart, patient is able to push up to 20 lb. Her pulling limitation was 16 lb.

The Jamar dynamometer grip test was also completed:

	Right hand	Left hand
1st setting	53 lb	32 lb
2nd setting	74 lb	68 lb
3rd setting	72 lb	68 lb
4th setting	61 lb	57 lb
5th setting	58 lb	53 lb

If patient is exerting maximum effort, the above grip strength measurements should represent a bell-shaped curve. The patient's grip strength for her age and sex should be 60 pounds taken on the second setting. This was achieved, and she had 74 lb of grip strength with the right hand. Patient's grip strength on the left hand, second setting, should be 53 lb; she had 68 lb of grip. This is within normal limits.

The Jamar pinch gauge was utilized for both the lateral and 3 jaw pinch prehension testing. The patient's right hand lateral pinch prehension strength was 13.5 lb. The left, for comparison, was 12.5 lb. The patient's lateral pinch prehension strength is within normal limits for the right hand. She is slightly weak on the left hand. Goals are 13 lb.

Right 3 jaw pinch prehension strength was 12.5 lb. The left for comparison was 16.5 lb. Patient is slightly weak on the right hand. The norm for her age and sex is 14 lb. Patient's bilateral upper extremity active range of motion and strength is within normal limits. Patient's fine motor coordination is within normal limits in bilateral hands.

Endurance: Patient was able to sustain a full 2-hour functional capacity evaluation. At the start of her treatment, her pain was a level 4. At the end of her evaluation, her pain increased to a level 8 on a scale of 0–10, with 10 being the highest. The patient complains of shooting pain in the ball of her right foot, and it shoots up through her calf, behind her knee, and she complains of pain to the right of her suture for the L4 lumbar laminectomy.

Summary: The patient's bilateral lift capacity is 3½ lb. She is only able to lift from thigh to her eye level. She cannot reach overhead. She also cannot reach below her thigh level. Unilateral lift is limited to 4½ lb. She is able to perform this at all levels except overhead reach. She is able to push 20 lb and pull 16 lb. Her grip strength is within normal limits for her age and sex. Her pinch strength is slightly weak in lateral pinch on the left and in right hand 3 jaw pinch prehension. She is able to walk independently. She is able to climb 16 steps. She also can descend 16 steps. She can walk, climb, squat, crouch, and crawl; however, she does need assist when going from standing to other planes. She is able to use an assist of a wall or a stationary object. She was able to climb two steps of a 6-foot ladder. She is not able to climb any further due to the pain in the ball of her foot. Her job requires manipulating small objects, to lifting approximately 20–100 lb during the day. Her greatest effort will be required in lifting a ladder. Her job is in the medium/heavy physical exertion category. Her tool box was left in the trunk of her car and she was instructed to pick up work tools weighing no more than 4½ pounds unilaterally.

Recommendations:

1. I recommend that patient start a work reconditioning program, starting with 3 times per week for 2 hours and working up to 5 days/week for a maximum of 8-hour sessions to increase her tolerance and endurance level for lifting bilaterally, unilaterally, and for increasing her ability to work overhead and to squat, crouch, and kneel without assist. She is also in a physical therapy program for increasing her muscle strength and increasing her pain-free motion.
2. I recommend that the patient utilize a safety rail for getting in and out of a bathtub to avoid any additional injury to her back.
3. I also recommend that the patient utilize a lumbar roll and obtain instruction in correct body mechanics to avoid strain to her lower back region.

4. I recommend an on-site job assessment to insure that the patient will be able to return to her former position as a contract electrician.

Prognosis: Fair+ to Good−. Recommend she follow up with her physician in 2 weeks. Patient/family education and a home program was issued.

Thank you, Dr. _____, for this referral.

Registered Occupational Therapist: ______________

EXAMPLE 5: ON-SITE OCCUPATIONAL THERAPY JOB ANALYSIS

Data Base:

Patient Name: ________________ Medical Record Number: ____________

Date: ____________

Diagnosis: Low Back Lumbar Sacral Sprain Date of Birth: ____________

Physician: ______________ Location: Steel Industry

Job Title: Material Handler

Insurance: Workers' Compensation

Supervisor or contact person: ________________________

Date of Injury: 2 weeks ago, 2-24-99

Past Medical History: None Reported Medications: None

Subjective

Job Summary: The patient operates a crane to receive stock, ship, and maintain flow of materials to and from storage areas and process operations. He is 41 years old and slightly obese. He utilizes an overhead crane and is responsible for bringing cores over to the plant handlers to have the cores sawed or cut. Tools and equipment also include a flatbed truck, scale, lifting devices, magnets, lubricating equipment, banding tool, measuring tapes, gauges, various sizes of stainless steel coils, bands, clips, and scrap material. He reports no past medical history and is ambulating with a cane. He refused to take medications per report.

Description of Tasks: The patient was able to bring coils from the hi-lo using the overhead crane. He was able to operate the pendant. He was able to place cardboard onto the floor for setting coils down. He was able to transport materials to and from the processing units. The patient was able to climb in and out of the tractor. He did have poor posture sitting. His knees were not higher than his hips while sitting. He was able to climb over an 8½″ high storage rack for the reels. He was able to ambulate utilizing his cane for an assist to shift the weight off of his back. The patient was able to hold the pendant in his right hand and utilize his left hand to align the crane to hook the steel coil. The patient was not able to raise the pendant to the cable. The pendant weighed approximately 50 pounds.

The patient was also not able to lift a 21″ diameter cardboard core weighing approximately 25 pounds and was not able to lift the 36″ long and 48″ long cardboard cores with bilateral upper extremities.

He was able to insure that proper completed identification tags were on all coils going in and out of the stock area. The patient was able to hook up or unhook for crane and change lifting devices as required. The patient was able to weigh in process material and record weight on designated forms.

He was able to insure that coils were properly banded for further moves, tighten or replace bands in coil cards, etc. He was able to unload and load coils and other materials from and onto the truck.

He was able to operate the tractor or crane to transport mill equipment, tools and maintenance and other operating supplies about the plant as necessary.

He was able to check, change and maintain proper oil, fuel, and water supply for tractor and change lift or towing devices as required. He was able to inspect, lubricate, clean and record condition of cranes operated. He is able to maintain a safe, clean, and orderly condition of the work environment. He can aid in the repair of maintenance work by making necessary lifts.

The patient is able to climb 14 steps and then another 10 steps. He was able to climb and descend utilizing the rail. He has a lumbar cushion that he places in his seat to operate the crane. The patient does have difficulty in the number four finishing crane job due to the stairwell being too narrow for patient's body. The patient needs to twist his lower back through the 17″ wide passageway to operate the crane cab. This forces the patient to put undue stress on his lower back in the lumbosacral region.

The number two finishing crane is a job that does not require the patient to lift the pendant. This job appears to be within the patient's physical capacity. The patient does have difficulty with lifting the pendant into the crane and with lifting cores for being cut. It appears that material hand number 419 is within the patient's capacity and ability.

Vocational Preparation: The job includes on-the-job training and a high school diploma.

Physical Demands:

Strength: The job is in the light physical exertion level, lifting hand tools and operating the tractor and crane. If, however, the pendant and the core job duties are added to this physical demand, the category would increase to a medium exertion level, requiring lifting a 50-pound pendant. The job requires frequent flexing and extending of the elbows for lifting and carrying objects.

Standing: The job may be performed standing. Standing tolerance requires up to 20 minutes. Patient leans on objects to take the stress off body parts.

Walking: Frequent walking is performed in the entire building. Patient had good ambulation.

Sitting: The job requires frequent sitting for an 8-hour and 10- to 12-hour shift. The patient must be able to sustain the sitting posture frequently, but the patient is able to sit for up to 2 hours. Then he walks, stands, stretches, and resumes the sitting position.

Lifting: The patient's job requires frequent lifting of tools such as small gauges, power tools, scales, gauges, lubricating equipment, bands, clips, scrap. No cane was used.

Carrying: The patient must be able to carry objects that require him to keep objects close to the body with the back in normal position. He was instructed in correct lifting and carrying techniques to avoid pressure to the lower back causing additional strain. The patient was able to ambulate 150 feet in the plant to reach another sit-down station (e.g., tractor cab). He walked at a moderate pace. The patient is able to push and pull. He had good body mechanics during this activity. The patient was able to pull levers forward for the hoyer lift. This is done in frequent and repetitive motions. No cane was used. The patient was able to climb and descend 15 steps.

Balancing: The patient was able to maintain his balance for climbing and descending the cab of the tractor holding on to the rail grab bar.

Stooping: The patient was able to stoop, crouch, and kneel from a standing position to pick up a pen. Good balance was observed. Stooping is not performed repeatedly on the job. Kneeling is minimal, crouching is minimal, crawling is minimal.

Reaching: The majority of the patient's work is performed in a total body range of motion. The patient works a lift from waist to shoulder level to the overhead position. Bending is frequent for picking up tools. Balance, stature is good. Driving must be performed for up to a 12-hour shift. Hand movements are frequent, turning, pulling, pushing, fingering, and twisting, which requires frequent elbow flexion and extension as well as wrist flexion and extension and gripping. Twisting is minimal.

Environmental Conditions: The majority of the patient's work is performed in a large plant area. Winter temperatures are comfortable due to heating. The summer months may cause the plant to be warm. The patient is not exposed to any severe weather conditions of heat or cold. Vibration is experienced from the levers that control the tractor cab.

Comments

This job requires light lifting, pushing, and pulling of objects; overhead work and standing is moderate; repetition in performing the job is frequent for bilateral upper extremities.

Recommendations

1. I recommend correct body mechanics so that the patient may stabilize his back and utilize the correct techniques of lifting and carrying to avoid undue pressure, and to ensure proper posture of the low back area.
2. I recommend that a crank be utilized if the pendant is required to be lifted into the tractor cab.
3. I recommend that the cores are brought over to the "plant handler" and that a "packaging" person cut the core. The number two finishing crane appears to be a part of the materials handler job. The patient is able to perform his job without lifting the pendant and by having the cores cut. It appears that other workers in the plant can cut cores before they reach the patient.
4. I recommend that the patient stay in a job capacity with the precautions of 20 pound weight restriction of lifting and no prolonged standing, stooping, or bending exists.
5. I recommend he follow up with his physician every 2 weeks. The cane is not required at work.

Thank you, Dr. _____, for this referral.
Respectfully,

Therapist: ________________________
Registered Occupational Therapist

EXAMPLE 6: OCCUPATIONAL THERAPY ON-SITE JOB ASSESSMENT: JOB SITE COACHING AND TREATMENT

Date of Evaluation: ___________

Patient Name: ______________________

Diagnosis: Bilateral Synovitis

Date of Injury: 3/2/99 prior injury date 8/11/97

Physician: Dr. _____, Occupational Medicine

Occupational Therapist: ______________________

Senior Benefits: ______________________

Analyst: ______________________

Location: Banking Industry

Supervisor: ______________________

Job Title: Senior Cash Letter Clerk

Internal Check Processing Department

Group-CLPA

First Shift Employee

Date of Birth: ___________

Insurance: Workers' Compensation

Job Summary: ______________________

Under guidance sort and distribute items from IBM processor. Check processed work for accuracy, locate problems, and prepare for distribution. Assist cash letter preparation clerks with problems.

Key Responsibilities

1. Receive computer listings, verify first and last batch items and printed cash letter by checking against computer printout, wrap bundles of work, and sort according to bank location. Verify the bundles are properly packaged prior to sending.

2. Guide cash letter preparation clerks with problems such as locating lost bundles and correcting misreads.
3. Check for misreads, missing items, and large dollar blocks and adjust totals as necessary.
4. Prepare general ledger entries for direct send work as well as prepare manual cash letters.
5. Verify all million dollar items in batch by cross-checking cash letter, computer printout, and batch items.
6. Insure receipts are maintained for outgoing cash letters.
7. Keep management informed of any problems that could result in missed dispatch deadlines.
8. Hand-deliver work to the Federal Reserve Detroit Branch or to Banking Garage for pickup by Bankers Dispatch.
9. Keypunch Federal Reserve cards as necessary.

Machines, Tools, and Equipment Used:

Burst machine
Computer paper and IBM processor
Strap machine
Plastic bags
IBM 424520 machine
Fold machine

Current Restrictions: Issued 3/4/99 by Hand Surgery Associates, P.C., Suite 101, 8560 Silvery Land, Dearborn Heights, Michigan 48127. "Continue present restrictions, 10 minute breaks, every one to two hours, cannot make boxes."

Subjective: The patient is 5 feet, 5 inches tall. She weighs 213 pounds. She is 58 years old. The patient is right hand dominant. She worked in 1993 through 1996 as a proof operator. Currently she works as a "check wrapper in the check processing department." She was initially out of work for 3 to 4 months. She presents with bilateral cock-up splints, which she wears on the job. The splints were prescribed by her occupational medicine physician. She sleeps at night in another set per report. The patient has seen many physicians for her condition. The EMG tests were negative on 1/7/99. X-ray reports are also negative. Bone scan is negative as reported by Dr. Slinger. Patient takes Naprosyn 500 mg two times a day.

The patient reports independence in activities of daily living. She states that since she has not worked the "proof operator job," she has improved in her physical capabilities. She reports that her chief complaint is in dropping objects with both hands. She states, "I have a new grandchild and I am afraid of dropping the baby." She reports enjoying her current job and supervisor. She states that she has not compared her electromyelogram (EMG) test results from previous years with her current results. She asks, "Do I need carpal tunnel release surgery to decrease my dropping things?" This therapist discussed conservative treatment techniques to prevent the need for surgery. Education was provided for correct posture and exercise at home and at work.

Objective: An on-site job analysis was performed by breaking the job into task performance segments, then analyzing each task according to the physical demands and frequency with which the demands are performed. Work station ergonomics and recommendations for redesign or adaptations and upper extremity wellness/posture was completed.

Hand Function: Active range of motion (AROM) is functional. Strength in the flexor carpi radialis muscle is 3+/5 bilateral hands. Her opposition, all digits for prehension strength is 3/5. Skin color is light and spotty, temperature is warm. No edema is apparent. She can performs small and large grasp; pulp, tip, and distal and proximal hook; and scoop functions. Strength in the left hand flexor carpi ulnaris is 4/5.

Pain Level: Pre- and postevaluation, with 0 being no pain and 10 being the highest, the patient reports her right hand at a pain level of 5 and her left hand at a level 4. Pain is over the flexor carpi radialis on the right and flexor carpi ulnaris on the left. She states that she drops objects frequently and that her pain radiates up to her elbow at times bilaterally. The patient complained of pain in bilateral wrist and hand dorsal surface. Skin is light in color possibly due to the stockinette pulled over both wrists used to line her cock-up splints. Patient reports numbness and tingling sensation in both hands.

Task/Work Performed:

1. Wrapping.
2. Cash letter (two times a day) preparation on table for delivery to banks.
3. Boxing and cash letter (checking for misreads and general ledger entries, computer printout, and batch items).

Description of Tasks Seen by Therapist:

1. Uses left hand to place bundle of checks into a box. Alternates left then right hand while wearing cock-up splints. The patient works while sitting.
2. Flips calculator tapes "to find the first check that is going to be in the next box."
3. Walks to another table 15 feet away to find boxes. Rubber bands the checks that do not fit in boxes when the boxes are full. Total number of boxes are 8 to 17. "Tuesday and Wednesdays are higher box days per report." The patient uses her right hand primarily to rubber band, but holds the bundle with her left hand.
4. Places tapes in the box on top with her right hand. She states, "I

cannot make boxes.'' She states that ''my work comes to me in boxes and this is okay in my department.'' She carries the box to the strapping machine using both hands.

5. She uses the strap machine to seal the box with four straps. The boxes may be 2 long boxes or 4 short boxes. I recommend that she strap no more than 4 boxes together to control the amount of stress applied on the tissues of bilateral hands. The total amount of stress is a product of the intensity, durations, and frequency. The patient reports that she has been abiding by the 4-box strapped-together process.
6. The patient performs the cash letter job totaling bundles and number of bundles for an APA number. She reports that she checks for misreads, missing items, and large dollar blocks and adjusts totals as necessary.
7. She examines 4 bundles at a time for a tracer number that correlates to the cash letter. She folds the document in two and rubber bands it using both hands. The left hand holds while the right hand folds and rubber bands. She uses size 32 rubber bands. Bundles may be 15 to 20 in number.
8. She places trays in the wall rack.
9. The patient states, ''I wrap on and off all day long, at 3:00 p.m. a courier comes in to pick up the work to deliver to the appropriate branch.'' She sorts according to bank location. She plastic bags 3 or 4 cash letter bundles. I recommend that the patient slide the bundles using both hands into the plastic bag on the table at waist level. Sliding 5 to 6 bundles is preferred over lifting. The 13-bundle bags were observed. Patient was able to use larger joints to slide the bundles into the plastic bags as instructed by the therapist. Body mechanics were reviewed.
10. Patient uses smaller plastic bags for 3 or 4 bundles. She labels and twists the bags for lift into the basket. The average number of bags is 10. She reports that ''the Denver bag is the heaviest, with 13 bundles.'' I recommended that the plastic bags also be slid off of the table into the baskets.
11. The patient reports that every 2 weeks she takes standard computer paper off of the IBM 424520 machine and pushes it into a burst machine with her left hand pushing the button and her right hand loading the machine. She was observed doing this. Then she places the paper into the fold machine. She uses both hands.
12. The patient reports, ''I usually perform 'wrapping' which might vary between putting in the trays for other workers on another shift, or putting in a plastic bag with a check letter.'' She performs a different job every 2 weeks pertaining to ''wrapping.'' I observed her wrapping

each tape with a divider tape match. She checked the first and last check of each bundle (pocket #22). She folds the tape across and down. She wraps around the check, stuffs the envelopes, and rubber bands. She used her left hand to hold and the right hand to stuff and rubber band. She used her right hand to initial the tapes using a pen.

Production: She reports loading 8 or 9 trays in a day. ''Tuesdays are the busy days per report'' (pocket #22). She rotates jobs every 2 weeks, which consists of pockets 14, 22, 15, 12, 13, 23, and 17. Pocket 20 is the burst machine. The other pockets, she states, ''are like the above duties as described.''

Physical Demands: The patient was able to perform all of the duties of her job requirements. Her work station is carpeted over concrete floor. The lifting appears to be in the 10 to 20 pound range with the weight lift being infrequent depending on the size of the Denver plastic bag and number of boxes. Reaching with arms/hands is frequent, up to 28 inches and repetitive. Bending is frequent with her back/waist and knees for lifting trays, bundles, and bags. She sits approximately 4 to 5 hours a day between standing and walking as needed to use the burst machine, folder, and strap machine. The job is in the light work level: lifting 20 pounds, with frequent lifting and carrying of objects weighing up to 10 pounds. It involves sitting most of the time with a degree of pushing and pulling of arms. Endurance for the evaluation was good.

Recommendations:

1. The patient should continue working full time in this job with unrestricted duty. However, I recommend that she continue wearing bilateral cock-up splints on the job to eliminate repetitive wrist flexion with active finger flexion and ulnar and radial deviation. The splints were checked for proper fit. The boxes should continue to be made up, as patient reports that ''her work comes to her in boxes.''
2. The patient should continue rotating between the pocket jobs every 2 weeks.
3. The patient should continue to strap no more than 4 boxes together to control the amount of stress applied on tissues of bilateral hands. I also recommend that she slide 5-bag bundles on the table using both hands rather than lifting. She should use larger joints to slide the ''13-bundle bags'' into the baskets, and use proper body mechanics as instructed.
4. The patient should take part in initial physical or occupational therapy treatment to decrease pain symptoms (ranging from dropping objects to numbness and weakness). In view of acute synovitis, passive exercise is used to maintain soft tissue mobility, as well as mechanical elasticity. Gentle controlled passive exercise will influence collagen formation without inflicting pain. It is used when joint mobilization

and stretching techniques are required to improve soft tissue mobility. The program will advance to medial nerve glide and controlled muscle strengthening.

Assessment: The patient appears to understand the importance of the above program. She tolerated the evaluation well.

Plan: Follow up with her physician every 2 weeks and be scheduled for physical or occupational therapy. A home program/patient education will be provided by the physical or occupational therapist.

Thank you for this referral.
Respectfully submitted by:

Therapist: ______________________
Registered Occupational Therapist

EXAMPLE 7: RETURN TO WORK CENTER JOB ANALYSIS

Name: ________________________ Date: July 13, 1997

Injury: Bilateral Epicondylitis Insurance: Workers' Compensation

Date of Onset: September, 1999—left elbow; June 23, 1998—right elbow

Therapist: ___________________ Date of Birth: ____________

Job Title: Janitor

Industry: Automotive Industry

Employer: Department of Facilities

Supervisor (Contact Person): ________________________

Title: Supervisor

Address: ________________________

Telephone: _________________

Job Summary

The *Dictionary of Occupational Titles* Job Summary Number is 382.664-020 (cataloged at the plant). Patient performs janitorial work of cleaning. This entails cleaning the automotive industry building, which is all on one floor. Patient has no stairs to climb.

Description of Tasks

Using bilateral upper extremities, broom, dustpan, power tools such as scrubber and other mechanical devices, (hydraulics) patient:

1. Empties all trash. Trash includes a total of 9 trash bins and changing trash bags within the 9 barrels.
2. Changes and lifts individual pieces of plastic from another trash disposal waste basket. There are a total of 10 dumpster-type barrels.
3. Sweeps floors.
4. Scrubs floors within the building using a mechanical scrubber.
5. Pumps in soap solution into the scrubber.
6. Uses mechanical hoyer to lift hoist to clean under hoist.
7. Pushes and pulls computer tables to clean underneath.
8. Picks up tires to place in dumpsters.

9. Pulls and pushes levers on scrubber; controls water release solution valves.
10. Locks garage doors.
11. Rings out and mops.
12. Cleans floors and commodes and sinks in the men's rest room.
13. Dusts, empties trash and ashtrays in the cafeteria; also lifts chairs and cleans tops of chairs.
14. Washes windows occasionally.
15. Cleans microwave occasionally.

Vocational Preparation

Includes on-the-job training and a high school diploma.

Physical Demands

1. Strength—heavy to very heavy—lifting 50 lbs. and sometimes up to 100 lbs. with frequent flexing and extending of elbows for lifting or carrying objects.
2. Standing. Job must be performed standing. Standing tolerance required up to 7 hours.
3. Walking. Frequent walking as work is performed in the entire building.
4. Sitting. None.
5. Lifting. Frequent lifting of heavy objects—tires, new car prep, empty boxes and plastic. Lifting tops to waste baskets and lifting chairs. Also lifts garbage trash bags, which may weigh up to 50 lbs.
6. Carrying. Must be able to carry cleaning products and trash. Patient carries trash from inside building to outside compressor.
7. Pushing. Patient must push garbage down in barrel. Patient must also push "super grease barrel soap pump" for loading the scrubber.
8. Pulling. Patient must pull lever forward for hoyer lift. Patient must also pull lever for scrubber. This is done frequently with repetitive motions.
9. Climbing. Minimal.
10. Balancing. During pushing or pulling of stools.
11. Stooping. Frequent use of dustpan and when mopping and removing and placing garbage bags.
12. Kneeling. When picking up trash from floor which may be minimal to moderate.
13. Crouching. Minimal to moderate when picking up trash from floor.
14. Crawling is minimal.
15. Reaching. Frequently during dusting. Majority of work is performed in a total body range of motion. Patient works and lifts from ankle

level to waist level to shoulder level to overhead position during cleaning.

16. Bending. Frequent, for picking up tools, picking up trash, and dusting.
17. Driving. None.
18. Hand movements. Frequent turning, pulling, pushing, fingering, and twisting, which requires frequent elbow flexion and extension as well as wrist flexion and extension and gripping.
19. Twisting. Frequent when turning the mop, broom, and scrubber.

Environmental Conditions

The majority of work is performed inside a large garage area. Winter temperatures are comfortable due to heating. The garage doors will open frequently during the day, but winter conditions appear comfortable. During the summer months, the air-conditioning will be occasionally shut off and the garage area becomes warm. Patient is not exposed to any severe weather conditions of heat or cold.

Noise and Vibration

There is extreme noise with exhaust, cars being started up. Also, horns are honked and there is the noise of the scrubber machinery and technical service institute mechanics prepping new cars. Vibration is experienced from the scrubber and from the levers that control the scrubber; also when patient uses the pump to pump in the soap solutions into the scrubber.

Hazards

There are carbon monoxide fumes, exhaust, and engine heat. There may also be gas, transmission fluid, and oils.

Atmospheric Conditions

1. Fumes
2. Odors
3. Dust
4. Gases
5. Poor ventilation

Machines, Tools, Equipment, and Work Aids

Job is performed using a variety of standard cleaning tools, scrubbers, mops, and hydraulics.

Comments

Job requires heavy lifting; pushing, pulling of objects. Overhead work and standing is extensive. Repetition in performing the job and vibration create possible risk of reinjury.

Recommendations

1. I recommend correct body mechanics so that patient may stabilize bilateral elbows (due to epicondylitis) and bend at the knees, using total body turning, with elbows stabilized at side to prevent repetitive elbow flex and extend.
2. I also recommend that patient utilize a lighter scrubber for performing the cleaning of floors. I feel that the levers on this current scrubber are resistive and place greater demand on elbow joints.
3. I further recommend that the patient attend therapy, 4–5 sessions, to receive further training in correct body mechanics since patient has a history of working for 25 years in janitorial work. The patient enjoys work and does not wish to be retained for other work in the plant.

Registered Occupational Therapist: ______________

CONCLUSION

The format of industrial rehabilitation records and reports are best arranged by listening carefully to and acting on what customers think they need, not what staff members think they need. The insurance carriers, physicians, and plant or company contacts all have program ownership. The types of reports and activities allow for multiple contacts with minimal resources. The plant or company is usually interested in knowing what part of a job the injured/ill worker can perform. Matching jobs to worker training and experience is the key. The process of work classification and occupational categories is identified in the Occupational Information Network (O*NET), which will replace the Dictionary of Occupational Titles. O*NET describes skills, manipulative tasks, and interests, knowledge, skills, and abilities from a cognitive viewpoint. Most data come directly from workers themselves in describing what they do, the skills they need, and the knowledge they use on the job. Physical demands are determined through on-site job analysis. Determining the essential functions of a job is done on a case-by-case basis. Real-life situations found in an individual job with an injured/ill worker must be evaluated. Work requirements in relation to job restriction must be measured appropriately. Functional capacity evaluation results help to clarify worker physical ability. Job hazards are categorized as to the frequency of activity and the likelihood of injury. On-site therapists, nurses, and physicians should never ask for suggestions if they do not plan to act on what they hear from the customer. Purchasing is not part of the daily operation of the medical facility once a supplier of health care is selected.

On-site occupational health core services include injury/illness treatment; new hire physicals; executive physicals; lab screening; drug screening; worksite nursing; work functional capacity evaluation; work reconditioning; on-site job analysis; occupational therapy; physical therapy; case management; wellness, prevention, and fitness; ergonomics; health education and risk screening; surveillance, and regulatory compliance. The program configuration may also include employee assistance programs, health risk appraisal, and overall health promotion. Referrals are sent to nutritionists, psychiatrists, psychologists, vocational counselors, and other health care providers as appropriate. The long-term commitment to the field of occupational health and industrial rehabilitation requires an understanding of how to rewrite or modify systems to reach the full potential for value.

Developing the on-site model requires increasingly more knowledge and skills to address the legal issues and to suggest solutions for job placement within the workers' physical and mental capability. Convenience and high-quality medical care not only reduces compensation costs but provides the worker with 24-hour care at work where 40% of their waking hours are spent. The costs of disability versus the costs of health care need to be measured for best practices.

Task forces are being developed across the country to develop community-based practice. The focus on wellness and prevention at the on-site setting supports the core values of occupational therapy and the marketplace. National commissions set up to explore competency by data collection and surveys collected by employers and other health care professionals take into account outcome and cost. While most therapists are expanding their scope of practice, registered occupational therapists seek daily to achieve functional outcome for patients. The use of high-tech standardized equipment may not be available on site in all plants based on the customer's willingness to pay for expensive pieces of equipment. Sharing of information and creating a tool to assist the employer assess a worker's capability is critical. A work capacity evaluation performed for a company on site without the use of sophisticated computerized work program equipment to evaluate, for example, the electrical trade or any other skilled trade can be summarized in relation to job tasks. Some plants or companies prefer ''not to have to calculate'' and don't understand how. Some industrial rehabilitation equipment correlates to the work duties and the environment. The union is asking health care clinicians to bring the medical information for job placement to the union membership in easy-to-understand terms. The author has purchased certain industrial rehabilitation equipment for her on-site locations in conjunction with the use of plant-supplied machinery and parts for job simulation (see Appendix 4).

Workers' compensation is the last frontier for managed care, but trained on-site physicians, nurses, and therapists can economically manage disability in a work/plant environment. Restoring injured/ill workers to full duty surrounded by the physical agents of manufacturing (e.g., vibratory tools and air guns) is gaining acceptance as we strive to bring health care on site.

APPENDIX 1: JOB DESCRIPTIONS

On-Site Plant Industrial Rehabilitation Physical Therapist

General Summary

Implements therapy at the patient's worksite, based on the patient's psychological factors and physical job demands. Recommendations are made for ergonomic redesign, extended restricted duty status, or medical retirements. Evaluative reports are used for independent medical examinations (IMEs) and litigated cases.

Principal Duties and Responsibilities

Works in conjunction with the plant medical personnel, job placement coordinators, workers' compensation representatives, union representatives, and plant manager in the planning of physical therapy treatment programs.

Reviews and interprets industrial rehabilitation prescribed physical therapy treatment programs; selects and administers appropriate therapy technique (pain management, work therapy, transitional work, work tolerance) designed to facilitate the maximum recovery of functional use of injured extremity. The purpose of the work therapy program includes promoting active use, and minimizing physical complications from cumulative trauma disorders through the implementation of control strategies to reduce and eliminate potential Occupational Safety and Health Administration (OSHA)–cited disorders in the work place.

Consistent with prescribed treatment and professional assessment of patient needs, increases the patient's strength and coordination, which facilitates adjustment to injury and return to work. Work tolerance exercises are divided into levels of resistance and physical exertion to improve prehension, lifting, climbing, crawling, bending, etc. Manual therapy techniques and modalities (e.g., heat, phonophoresis, massage, electricity, ultrasound, whirlpool) may be applied.

Documents all care given to patients. Confers with plant physician and nursing staff regarding patient progress and modifies treatment plans as appropriate.

Instructs patient and/or families in appropriate home programs, fits patients with crutches, braces, and other walking devices. Examines patients for muscular skeletal impairment, and evaluates ability to perform proposed duties with respect to company needs and ADA requirements.

May conduct or participate in in-service education programs in physical therapy for industrial rehabilitation staff, students, and other plant employees.

May lead or direct the activities of physical therapy assistants, technicians, aides, athletic trainers, attendants and students, ensuring that department policies and procedures are enforced and that quality patient care is provided.

Consults with company/plant disability management to resolve employee/patient lost work time. Discusses specific return-to-work guidelines and modified programs for injured workers.

Knowledge, Skills, and Abilities Required

The job requires a specific level of technical knowledge of OSHA, JCAHO, Bureau of Workers' Compensation, and company plant reporting requirements. Knowledge of union laws and collective bargaining pertaining to ADA, restricted work, on-site job analysis, ergonomics, and workplace safety standards is required. Technical knowledge of physiology and application of various treatment interventions in order to relieve pain and/or restore physiological functions is needed. This level of knowledge is generally obtained through completion of an accredited approved American Physical Therapy Association (APTA) program with a bachelor's or master's degree and registration/licensing in the State of Michigan. Familiarity with industrial medicine practices (i.e., self-insured employers and worker tax-free wage-loss benefits that meet and exceed regular wages and supervisors who do not believe in allowing injured workers to return to work may change the course of treatment) and scope-of-practice guidelines as determined by the APTA is necessary.

Determines fitness for duty based on knowledge of patient's condition and reports from employee's physician.

The job requires certain physical abilities in order to lift and position patients and to perform treatment modalities. The physical demand characteristics of work are in the very heavy level; weight lifted may be in excess of 100 pounds, frequency of lift is infrequent at 100 pounds and frequent at 50 pounds to 100 pounds; walking and carrying are 3.5 mph with 50 pounds or more load; typical energy required is 75–120 METS.

Working Conditions

Works in a plant patient care environment; care and alertness are required when working with and around physical therapy and industrial plant equipment.

On-Site Industrial Occupational Therapist

General Summary

Implements therapy at the patient's worksite, based on the patient's psychological factors and physical job demands. Performs on-site job analysis by breaking down a particular job into its essential duties and physical demand characteristics. Exertion levels, weight lifted, frequency of lifts, walking and carrying loads, and typical energy required are evaluated. Ergonomic designs of equipment, facilities, work methods, and tools are evaluated and/or redesigned. Education and training of employees at the workplace in prevention of illness and injury and safety of the work station create an appropriate conditioning program to minimize lost work time. Delivers job coaching and proper conditioning of patients in a transitional work placement. Recommendations are made for extended restricted duty status or medical retirements. Evaluative reports are used for independent medical exams (IMEs) and litigated cases.

Principal Duties and Responsibilities

Evaluates industrial jobs that involve repetitive hand, wrist, elbow, and back and whole body movements and related tool use.

Identifies cumulative trauma of the upper extremities that includes but is not limited to tendinitis, bursitis, synovitis, ganglion cysts, trigger finger, DeQuervains tenosynovitis, and carpal tunnel syndrome.

Consults with plant medical personnel, job placement coordinators, workers' compensation representatives, union representatives, and plant manager in the proper posturing (e.g., keyboard height, overhead work).

Performs job coaching for stretching, starting work, rotation, and breaks from repetitive work. Supplies adaptive equipment (e.g., articulated keyboard arms, cushions, hydraulics, antivibration gloves, arm rests, holders, pendants, ramps, lifts, etc.).

Designs ergonomic buttons, surfaces, adjustable footrests, and other portable or built into the workplace adaptations for prolonged sitting, standing, squatting, crawling, bending, climbing, etc.

Evaluates and measures leaning forward and reach distances in the work area, sitting with the body twisted (due to inadequate leg clearance) or improper low back support or use contributing to pain and other symptoms in the neck or back.

Consistent with prescribed treatment and professional evaluation and assessment of patient's needs, increases the patient's strength and coordination and decreases pain and edema. Custom fabricates splints, performs hand therapy treatment programs, and increases patient's strength and endurance in relation to work duties.

Documents all care given to patients, confers with plant physician and nursing, and modifies treatment plans as appropriate.

Instructs patients and/or families in appropriate home programs.

May participate in in-service education programs in occupational therapy for industrial rehabilitation staff, assistants, students and other plant, benefits, and ergonomic employees.

May lead or direct the activities of other physical and occupational therapists, athletic trainers, technicians, assistants, students, ensuring that the department policies and procedures are enforced and that quality patient care is provided.

Consults with company/plant disability management to resolve employee/patient lost work time. Discusses specific return-to-work guidelines and modified programs for injured workers.

Knowledge, Skills, and Abilities Required

The job requires specific level of technical knowledge of OSHA, JCAHO, Bureau of Workers' Compensation, and company/plant reporting requirements. Knowledge of union laws and collective bargaining pertaining to the ADA, restricted work, on-site job assessment, ergonomics, and workplace safety standards is necessary. Technical knowledge of usual and customary job duties, workplace machines, tools and equipment used, restricted work, transitional work, physical demands and upper extremity/hand function is required. Production standards at the workplace and functional capacity evaluations will be summarized for employers. The U.S. Department of Labors' Physical Demand Characteristics of Work (Dictionary of Occupational Titles) will be referenced along with the O*NET database on a case-by-case basis. This level of knowledge is generally obtained through completion of an accredited approved A.O.T.A. program. Board Certification by the NOTCB and a license to practice in Michigan is required. Follows scope of practice guidelines as determined by A.O.T.A.

Determines fitness for duty based on knowledge of patient's condition and reports from employees physician.

The job requires certain physical abilities in order to lift and position patients and to perform treatment modalities. The physical demand characteristics of work are in the very heavy level; weight lifted may be in excess of 100 pounds, frequency of lift is infrequent at 100 pounds and frequent at 50 to 100 pounds; walking and carrying are 3.5 mph with 50 pounds or more load; typical energy required is 75–120 METS.

Working Conditions

Works in a plant patient care environment; care and alertness are required when working with and around occupational therapy and industrial plant equipment.

On-Site Industrial Certified Athletic Trainer

General Summary

1. The athletic trainer may be assigned to responsibilities in an industrial setting (working with employees who are at work and enhancing their physical fitness and knowledge of exercise) or to responsibilities in a physical therapy and occupational therapy work setting.

2. In an industrial setting the athletic trainer must work with employees and a licensed physician, monitoring the employee within the laws governing the practice of athletic training.
3. Under supervision of a physical therapist the athletic trainer is responsible for the implementation of the physical therapist's treatment plan for patients, using accepted and established forms of therapeutic exercise and modalities. Prepares all required documentation and reports, provides supervision (but not delegation) of a physical therapy aide or equivalent, and assists with and participates in department educational and quality assessment programs.

Principal Duties and Responsibilities

1. Administers flexibility assessment to employees in an industrial plant setting in accordance with ACSM guidelines for exercise, testing, prescriptions, and American Heart Association exercise standards. Implements facility informed consent process to users of the fitness/wellness facility. Advises all exercise candidates of the benefits and risks of participation, testing, physical activity and advise them that participation is voluntary in nature. Sets up and administers preventative programs to target the ''well'' population. Classes offered by the athletic trainer can include: Heartsmart/nutrition, weight reduction, sports medicine (body composition and prevention of athletic injuries), lifestyle modification, aerobics and exercise classes. Educational materials and research articles will be provided as indicated.
2. Renders therapeutic treatment (following evaluating physical therapist's and/or occupational therapist's plan) to patients with various injuries and diseases in accordance with accepted and established standards.
3. Fully documents and reports patient treatment programs, progress, and discharge plans as outlined in department policies and procedures. The supervising physical therapist shall monitor and co-sign the athletic trainer's documentation as part of their supervision when the athletic trainer is providing physical therapy services. The same is true for occupational therapy (e.g., transitional work program).
4. Confers with referring and consulting physicians, nursing personnel, and other health professionals regarding patient's progress and status on ongoing recommendations.
5. Assists with discharge planning (consulting with evaluating physical therapist) by conferring with physician with recommendations for follow-up therapy and/or equipment needs and instructing patient and/or family in home programs as needed.

6. Advises department supervisor and secretary in regard to patient load and patient scheduling and treatment required.
7. Accepts patients from physical therapists. Follows treatment plan and regularly discusses initial evaluation, ongoing assessment and change in patient status/progress, treatment progression and/or changes indicated. Seeks out the evaluating physical therapist when reevaluation is needed and/or sets new problems and goals and changes treatment plan as needed, after consulting with and obtaining the physical therapist's signature. Consults with evaluating physical therapist for discharge planning as needed and obtaining signature for discharge summary. The physical therapist is on site at all times during patient care.
8. Where permitted by law, the athletic trainer will conduct routine operation functions, including supervision (but not delegation) of a physical therapy aide or equivalent.
9. Assists with and participates in department in-service education programs, maintains professional growth and development by attending continuing education seminars and presenting services to department, and attends and participates in regularly scheduled department meetings.
10. Assists the department supervisor and director in the department quality assessment and accreditation program as requested.
11. Takes on special projects and programs as requested.
12. While assigned to on-site industrial rehabilitation setting, the athletic trainer is responsible for the prevention of industrial injuries; recognition and evaluation of industrial injuries; medical referral; rehabilitation of industrial injuries; organization and administration of training programs; and education and counseling of employees.

Job-Related Interpersonal Contacts

Works in conjunction with the plant medical personnel, job placement coordinators, workers' compensation representatives, the union, and plant manager as well as other interdisciplinary health professionals within the rehabilitation services department, other industrial departments, and contract sites related to industrial rehabilitation. These will include physicians, patients, rehabilitation professionals, students, volunteers, employees and administrators, athletic training education representatives, and medical supplies and sales representatives.

Working Conditions/Environment

Works in an industrial setting, patient care environment; care and alertness are required when working with and around physical and occupational therapy and industrial plant equipment. The athletic trainer will be in contact with few toxic substances and will need to exercise precautions secondary to patients and industrial worker's medical conditions.

Job Requirements

A. *Education*

1. Minimum of a bachelors degree in physical education and related health sciences from an accredited program for athletic training.
2. Certification by the National Athletic Trainers Association (NATA).
3. Preferred/Additional—continuing education courses in athletic training and physical therapy field.

B. *Experience*

1. Minimum—general outpatient orthopedic clinic experience during student internship and a minimum of 1,000 hours of practical experience working as an athletic trainer at the college level.
2. Preferred/additional—prior experience in the field of physical therapy, athletic training, and industrial rehabilitation.

C. *License*

License or registration not currently required in State of Michigan. Prefer current or of state license registration.

D. *Skills and Abilities*

1. Knowledge of athletic training practice standards, policies, and procedures in accordance with the NATA, the Michigan Physical Therapy Association (MPTA), and other regulatory groups.
2. Knowledge of physical therapy practices, standards policies and procedures in accordance with the American Physical Therapy Association (APTA), the MPTA, and other regulatory groups (e.g., A.O.T.A.).
3. Good oral and written English communication skills.
4. Ability to work both independently, assuming responsibility and initiative within the limits of authority, and as a team member with other health professionals.
5. Ability to communicate effectively with good interpersonal communication skills and to work productively and cooperatively with various personalities (both coworkers and patients).
6. Ability to perform physically demanding work requiring frequent lifting, pushing, pulling, reaching, kneeling, plus prolonged sitting and standing.

7. Ability to safely perform athletic training and assist in physical therapy treatment and/or occupational therapy treatment in regards to patient safety, proper body mechanics, and use of universal precautions.
8. Ability to work standard 40-hour week. Other hours may be requested as needed.

APPENDIX 2: POLICIES AND PROCEDURES AND THERAPY GUIDELINES

Protection of Clinical Records

Policy Statement and Interpretation

Industrial rehabilitation is committed to maintaining clinical records on all patients. The clinical records are accessible and systematically organized according to name of patient to facilitate chart retrieval and compilation of information. Industrial rehabilitation recognizes the confidentiality of the record and provides safeguards against loss, destruction, and/or unauthorized use.

Procedure

All employees at hire date and quarterly will be familiar with the procedure governing the use and removal of medical charts to safeguard against loss, destruction, unauthorized use of records and computerized information, and release of information.

The medical record will contain sufficient information to clearly identify patient, diagnosis(es), and treatment and will document the results of treatment accurately. All records will contain general data (e.g., assessment, plan of care, services provided, identification data, medical history, physical exams, observations and progress notes, clinic, finding and discharge summary, with final diagnosis and prognosis).

Current and discharged patient records are filed or archived in sequential order at the industrial rehabilitation clinic's on-site company location. The clinical administrator and supervisor will be able to easily locate charts, and all medical staff will be familiar with automated electronic information systems retrieval.

Clinical records are retained for 5 years after the date of discharge or, in the case of a minor, 3 years after the patient becomes of age under state law or 5 years after the date of discharge, whichever is longer. No part of the record may be destroyed.

The patient will sign Form C.L.1 authorizing release of information in the medical record. The patient's written consent is required for release of information not otherwise authorized by law.

Industrial rehabilitation prohibits the use, or removal, of records from the rehabilitation facility premises, except in the transportation of the medical chart to another industrial rehabilitation facility for review (e.g., quality assurance audit, discharge planning meeting, or patient care meeting). The clinic administrator and supervisor will secure the name of the patient whose chart is en route to, or is maintained at, another rehabilitation site. A fireproof case must be used if transporting records off site.

Industrial Rehabilitation Referral Policy

Industrial rehabilitation accepts patients for treatment as follows:

1. All patients are required to have orders for treatment prior to being seen in industrial rehabilitation.
2. Referrals and orders are accepted from:
   a. M.D.
   b. D.O.
   c. D.D.S.
   d. D.P.M.
3. Referrals and orders are accepted from the physician in written form. An original signature is preferable, although a stamped signature is acceptable if we receive a letter from the physician stating that this is how the physician handles referrals.
4. In case of direct patient inquiry, the potential patient should be referred to his/her physician to obtain the required order, after which time an appointment can be scheduled.
5. No discrimination will be made by race, color, age, sex, or national origin in accepting referred patients, use of facilities, equipment, or personnel to provide treatment.

Referral

1. Patients in need of physical and occupational therapy services are accepted for treatment only on order of a state licensed physician.
2. For each patient there is a written plan of care established, and this is periodically reviewed by the physician.
3. The organization has a physician on call and/or access to an emergency medical service to furnish necessary medical care in case of emergency.

4. The following is made available to the rehabilitation staff prior to or at the time of initiation of treatment:
   a. The patient's significant past history
   b. Current medical findings
   c. Diagnosis
   d. Physician's orders
   e. Contraindications
   f. Patient and/or guardian awareness of diagnosis or prognosis
   g. Where appropriate, summary of treatments rendered and results achieved during previous periods of physical and occupational therapy
5. Plan of care. A written plan of care is established for each patient by the physician and/or therapist, which indicates modalities and specifies the type, amount, frequency, and duration of physical or occupational therapy services. Where appropriate, the plan is developed in consultation between therapist and the physician.
6. At the time of the referral and, given that the patient is accepted as a treatment candidate, the treating therapist will contact the job placement coordinator and case manager informing him/her of the referral so he/she might do the necessary workplace evaluation in conjunction with the interdisciplinary team.
7. Industrial rehabilitation will be responsible for sending progress documentation to the physician if he/she has not seen the patient within a 30-day period.
8. The treating physician is promptly notified of any changes in the patient's condition. If changes are required in the plan of care, such changes will be supported by a new prescription specifically documenting the needed changes in modalities and/or exercise techniques rendered, etc.
9. When a verbal order is received from the physician, the order is written down, initialed, and dated by the therapist with a verbal order notation. A signed prescription from the physician to substantiate a verbal order must be acquired.
10. All patients will need a new referral in the following instances:
    a. Cardiac or respiratory arrest
    b. Admission to a hospital
11. When the progress in therapy is impeded, or when the patient no longer benefits from further treatment, the therapist notifies the physician. If the physician agrees, the patient is discontinued from treatment.

12. At time of discharge, a complete summary of the patient's progress, attainment of goals, from the initial treatment to the time of discharge, is documented and summarized.

Clinic Record Documentation

Upon referral, an evaluation and treatment plan will be completed by a physical and/or occupational therapist. The documentation of evaluation, treatment progress, discharge, and follow-up summaries will be maintained and submitted to the referring physician and to third-party payers as required or requested.

Appropriate documentation will be kept on all patients. Every sixth treatment—or sooner, as required—a written report will be sent to the referring physician summarizing the patient's progress towards establishing goals.

Documentation of patient's visits and progress will be sent to third-party payers as required. Written documents will include the following:

1. Patient's name, address, identification number, and referring physician
2. Treatment diagnosis, date of onset, and precautions or problems relating to diagnosis
3. Initial evaluation
4. Estimated duration of treatment, including frequency of these treatments
5. Program plan and interim evaluations
6. Home programs, restorative equipment, discharge and follow-up plans, as appropriate
7. Signature of the attending therapist

Infection Control

The patients who are referred to the industrial rehabilitation clinic fall into two general categories:

1. Patient's skin integrity is not compromised.
2. Patient's skin integrity is compromised.

Group 2 will necessitate adherence to the following procedures to control cross-contamination.

I. Movement of patients within the department
  A. Skin and wound infection
    1. Where drainage is moderate and can be contained and kept dry with heavy dressing, cover wound with a clean towel and transport from hydrotherapy room or vice versa.

2. Where draining is excessive and cannot be contained by heavy dressing, cover wound with clean towel and overlay plastic and transport from the hydrotherapy area to treatment room.

B. Burned Patients

1. Cover with sterile dressing prior to transporting to and from hydrotherapy area.

II. Handling patients with draining wounds

A. Schedule these patients at the end of the treatment period

B. Special for disposal of contaminated dressing

1. Dressing will be removed by therapist or such aide as he or she designates who has had instruction in dressing techniques.

2. Use of gloves is required for ANY open wound.

3. All contaminated dressing materials will immediately be placed along with examination gloves in a plastic bag liner and double bagged. The bag will be marked TO BE INCINERATED OR REMOVED. Bloodborne pathogens and hazardous waste removal safety policies and procedures will be followed at each plant or company location.

Massage

I. Physiological effects

A. On circulation of blood

1. General circulation of blood—no significant effect
2. Local circulation of blood—a temporary increase
3. Venous return—some assistance

B. On circulation of lymph

1. Considerable increase due to milking effect
2. Reduces lymphedema

C. On nervous system

1. Sedative effect—mechanism unknown
2. No effect on nerve tissue, e.g., useless in denervation

D. On muscle tissue

1. No effect on muscle strength
2. Due to circulatory effects—improves nutrition

E. On skin

1. Helps free adhesions
2. Assists function of sebaceous glands and sweat glands

- **F.** On adipose tissue
  No effect on fat
- **G.** On bone
  No effect
- **H.** Psychological effects

II. Components of massage

- **A.** Effleurage (stroking)
  - **1.** Superficial and deep
  - **2.** Centripetal direction
  - **3.** Gliding over skin with the hands
  - **4.** Pressure is firm but gentle and varies
- **B.** Petrissage
  - **1.** Kneading—picking up movement with vertical compression
  - **2.** Friction—the part of the hand or fingers being used is kept in contact with the skin, and the superficial tissues are moved over the deeper underlying ones; loosens the underlying tissues
  - **3.** Various types of deep pressure strokes

APPENDIX 3: EMPLOYEE CORRESPONDENCE ON THE TEMPORARY MODIFIED WORK PROCESS

To: All Employees Date: June 9, 1999

Subject: Temporary Modified Work Process (TMWP)

Objective

To provide ''temporary'' modification of an employee's occupation while on restricted or limited duty status during a course of treatment, physical and occupational therapy, or work reconditioning program designed to return the employee to the full scope of his or her occupation.

Parameters

1. Qualification for TMWP shall be determined on a case-by-case basis and approved by the on-site occupational medicine physician, department supervisor, and the industrial relations department.
   A. Where the on-site occupational medicine physician concludes that TMWP is medically appropriate, the physician shall discuss the restrictions or limitations with the industrial relations department and the supervisor of the employee's department.
   B. The supervisor shall evaluate the department's ability to return the employee to a modified job in his or her department. This will include an evaluation of the effect of restrictions or limitations on 1) the employee's ability to perform the job satisfactorily, 2) the productivity of the operation, 3) the ability to shift work or modify work to accommodate the restrictions, and 4) the probability of injury.
2. The duration of TMWP shall be limited to a period of 4 weeks, which can be extended with approval of the attending physician by the on-site occupational medicine physician, department supervisor, and the industrial relations department. An extension of up to 4 additional weeks *must* be based on a reasonable assumption that the extended

period will result in return to the full scope of the employee's occupation.

3. Treatment, occupational or physical therapy, or physician appointments shall be scheduled during off shifts.
4. In the event the employee does not qualify for TMWP, nothing in the process will preclude the employee or the employer from exercising their rights under the labor agreement or any applicable state or federal laws.
5. TMWP shall be available under the same parameters, on a voluntary basis, for nonoccupational disabilities.

APPENDIX 4: ON-SITE REHABILITATION EQUIPMENT LIST

Volometer
Velcro
Heat gun
Adhesive felt
Terry cushion
Stockinette and coband
D-ring straps
Finger loop and sling and hooks with adhesive
Splint pan
Tennis elbow support/Neoprene sleeve
Exercise theraputty (all grades)
Finger block
Theraband (all grades)
Foam and action hand exerciser Band power hand grip (resistance varies)
Thumbciser
Dystrophile (progressive stress loading device)
Cuff weight rack with cuff weights (1–25 pounds)
Hand table and stool
Modalities (whirlpool, paraffin, electrical stimulation, ultrasound, iontophoresis, and phonophoresis)

Thermoplastics/perforated
Straps
Moleskin
Contour foam
Heel and elbow protectors
Velfoam
Outrigger, steel spring wire, slotted pulleys
Scissors, punch pliers
Slings
Dexterciser
9-hole peg board
Wrist, forearm rope wrap/roller
Digiflex and color-coded rubber hand exercisers
Deluxe pipe tree
Overhead arm pulley (door bracket)
Dumbbells (assorted up to 50 pounds)
Prefabricated finger, wrist and forearm splints (e.g., reverse knuckle bender, stax finger splints, finger flexion glove, buddy straps, spring PIP extension and flexion splints)
Hydrocollator (with cervical/lumbar hot packs)
Soft-tissue mobilization prelim-27
Tissue massage tool to decrease

Scar massage lotion and vibrator
Sensory nerve retraining kit
Total arm elevator (foam) to reduce edema
Latex-free gloves and dressings
Jamar dynamometer
Two-point discrimination
Force measurement gauge
Goniometers

Pegboard
Stopwatch and 60-minute timer
Tapemeasure and finger circumference gauge
Plastic skeleton model

Ergonomic products

Wrist savers, supports for chairs
Foot rest
Sorbothane work gloves
Push/Pull cart
Baltimore tool evaluation (BTE)
Floor mats
West system

4 × 6 exercise mats

Blankenship boxes
Valpar 9
Wooden sled
10-foot ladder
Typewriter
Bolt board
Hand truck
Table with chart rack
Spine dynamometer
Velcro checker board
Square wood pegboard
D.L.M. (consumer sequence cards)
Wooden chairs with grey seats

adhesions

Air splints, edema control gloves and sleeves

Semmes-Weinstein monofilaments
Pinch gauge
Shoulder, wrist, and finger
Minnesota manual dexterity, Purdue test set

Lafayette/grooved pegboard
Hand and wrist anatomical wall charts
Educational booklets and home exercise programs
Articulating drawers, keyboard, and mouse
Back lumbar rolls
Writing instruments and grips
Toolbox and weight scale
Lift task and crates
Work cube
Valpar system
Cybex UBE (upper cycle and seat bench)
W.E.S.T. lifting bucket system/ work cube
W.E.S.T. bus bench
Valpar 8
3-foot refrigerator
6-foot ladder
Roltztool
Milk cases
Pneumatic wood top table
6-foot table with Valpar 8
B.T.E. tool rack
Circular wood pegboards
Set multicolor stacking cones
D.L.M. (How to solve 1-2-3 step story problems)
Slide projector

Back school slide programs

O'Conner tweezer dexterity test
Wood saw
25-foot ruler
Health O'Meter scale
Micro fit pressure gauge

VHS camcorder (job site evaluation)
Set of hand tools and tool box
10-pound weight plates
Blood pressure unit
Set of Chatillion pressure gauges
Cervical and lumbar traction table
Mat tables
Baps board
Traction table
Cybex ab bench
Cybex roman chair
Cybex leg press
Cybex adductor
Cybex arm extension
Cybex overhead press
Cybex chest press
Cybex cable cross station
Cybex Upright Bike
Versaclimber
Track Treadmill
Stairmaster
Video and television
Locked files for records
Laundry service
Telephones
Printers
Dictaphone
Handicapped restroom
Lockers, break room
Drinking fountain

Clinic sink
Clean and soiled laundry rooms

Formica table (microwave/coffee maker)
Garbage cans
Hard hat
Hansen weight hook (50 pound)
Free weight (50 pound)
35MM camera (job site evaluation)
Talking scale (job site evaluation)

Electric screwdriver with tools
12″ Blankenship boxes
Pulse meter
Lumbar MedX
Plinths
Therapy balls
Acron Hi-lo table
Cybex ab crunch
Cybex back extension
Cybex rotary calve
Cybex wrist/forearm
Cybex abductor
Cybex arm curl
Cybex fly
Cybex chin/dip
Cybex semi-recumbent bike
Cybex upper body ergometer
Stairmaster FreeClimber
Concept II Row
Treadmill with EKG
Office space
Handicapped parking
Faxes
Beepers
Computers
Transcription service
Cabinets
Waiting room with table and chairs
Fire extinguishers and evacuation map
Cubicle containers and booths
Storage

6

Americans with Disabilities Act: Compliance, Worksite Ergonomics, On-Site Job Analysis, and Preplacement Coordination—How They Interrelate

An on-site job analysis involves the process of breaking down a particular job into its essential functions or parts in accordance with the Americans with Disabilities Act (ADA) requirements. A job analysis helps reassure supervisors about the capabilities of an employee with a disability. An on-site job analysis videotaped in a plant area to identify ergonomic problems and solutions will impact a larger population of workers who perform the same job duties. (Appendix 1 contains a sample report from a videotaped automobile company ergonomic job analysis).

Registered occupational therapists were able to report their observations upon seeing employees building a prototype car. A stockroom and a wood shop were also evaluated to improve on and/or prevent injury rates. Plants, manufacturing, and service industries have good intentions. Technological innovations may be promising, but, some do not work. Any solution to problems in the workplace needs to be dealt with proactively. Exposures to the variables obscure the true process because people differ in size and height, coordination, strength, range of motion, and cognition. Top management commitment and support of the workforce is essential in dealing with the pervasive effects of technological change. In some instances, however, top management commitment seems to hinder the innovation processes. This can be experienced when top management force innovation on an unwilling organization or when bureaucracy champions technology that can do irreparable damage.

The on-site job analysis can focus on key human resources, whose job duties or tasks may reinforce desired changes. The employers manufacturing or service mission and strategy are sensitive. The changes that allow the organization to do better and safer work more easily can be expensive or relatively cheap. Workers, group leaders, production/operations managers, and corporate staff should ask themselves, ''What is the most important thing for this company to be doing, and how do my job duties complement the goal?''

The term ''ergonomics'' comes from two Greek words *ergo*, meaning work, and *Nomos*, meaning law. Ergonomics refers to the ''laws of work'' or the way work should be. It involves the design of tools, equipment, and the environment to properly fit the individual using them in order to ensure work safety and efficiency. Many work-related injuries and physical complaints can be prevented by properly designing the workplace. This applies to the service industry and plant environments as well as the office setting. An improperly designed workplace can cause physical stresses on various parts of the body, resulting in pain and other symptoms that result in low productivity, injury, and even absenteeism. For office workers who spend a great deal of time seated at a desk or computer station, physical discomfort due to improper workplace design can last longer than the 9-to-5 work day. People frequently take home much of the fatigue and discomfort they experience, but they could leave it at work by following certain suggestions.

Wrist problems have long been associated with jobs that involve repetitive hand and wrist movements and tool use. Such problems are known as cumulative trauma of the upper extremities and include tendinitis, bursitis, tenosynovitis, ganglion cysts, carpal tunnel syndrome, and other nerve impingements. Workers should be educated as to the following ''do and don't'' techniques:

1. Don't work in a posture that causes stress to your hands and wrists and increases pressure on tendons and nerves. Check yourself when sitting at a keyboard. Do your wrists bend downward when you key? Do they rest on a hard or sharp surface such as a desk or desk edge?

Do keep your wrists held in a neutral position (0–30 degrees of wrist extension).

Do fit your desk with equipment that allows comfortable use of keyboards, such as a wrist rest to help reduce pressure and improve wrist posture and comfort. Also, use an articulated keyboard arm—one that is mounted in the drawer area, pulled out, and adjusted to the preferred height.

Don't place your keyboard on the desktop. Often the edge of the desk may press on the wrists or forearm, causing irritation over prolonged periods. A keyboard on a typical office desk may be acceptable as long as the keyboard sessions are occasional or infrequent and short in duration. It is not recommended if used regularly and for long periods of time.

Do adjust the keyboard to a height that allows the hands to be in line with the forearms. A keyboard that is too high or low can force the wrist into flexion or extension, which can cause problems if performed regularly for extended periods of time.

Do perform hand-care exercises that can stretch and warm up the hands before starting work or give the hands a break from repetitive work. For example, "hand-pumping" exercises can be done by making a fist and opening the hand. Hold each motion for a count of five as if pumping up a blood pressure cuff. Be sure to keep your hand elevated above the heart level. Perform three sets of 10 repetitions.

2. Neck, shoulder, and upper arm problems also can be caused by a work area that is too high or low or from the worker having to slump or lean forward to either reach the keyboard and work area or clearly see the monitor or hard copy.

Make sure there is adequate leg clearance under the work table. Lack of leg space can contribute to being too far from the work space.

Make sure the keyboard is at a proper height. Knees should be slightly higher than the hips.

3. Position the monitor or hard copy appropriately to avoid excessive head inclination or leaning. The placement of monitors depends on the size of the screen and characters. The smaller the screen, the closer it needs to be.

Do use a document holder—particularly those with an articulated arm—to hold copy upright and at the same orientation and distance from the eyes as the screen. This will help reduce excessive eye and head movement and inclination.

4. Leaning forward in order to reach the work area, sitting with the body twisted due to inadequate leg clearance, or sitting without proper low back support can all contribute to pain and other symptoms in the low back.

Remember that there is no such thing as an "ergonomic chair." The chair should be fitted to the person based on his or her height and size.

Have an adjustable chair. Chair height, back-rest height, and horizontal movements should all be adjustable.

5. Simple brief stretches throughout the day will help the body recover from the demands of work. Muscle fatigue is due to a buildup of waste products that develop in working muscle, combined with an inability

> of oxygen to get into the muscle. Stretching promotes a ''cleansing'' process by flushing out waste products and improving the flow of blood and oxygen into the muscle. Literally, stretching helps muscles to ''breathe'' and allows them to work.

Don't forget to stretch. Stretching:

reduces muscle fatigue
reduces muscle discomfort
makes it easier to move
balances the body

Do safe work-site stretching exercises for the legs and back, including back rotations, toe touches, side bends, and groin, hamstring, gluteal, and hip-flexor stretches.

Do hand-care exercises including fist, claw hand, and touching the thumb to the base of the fifth finger.

Education and training of employees in the prevention of illness and injury will carry over to behavioral modification of lifestyle. Employees can be observed directly at their work station and pulled off the job if a poor posture or improper body mechanics is evaluated by a board-certified occupational therapist, physician, or ergonomist. In the long run, performance solutions facilitate operations improvement, lower financial risk, reward people's performance, and demonstrate change in management and strategy implementation.

Specialists in workplace design or modification have health care knowledge allowing them to apply their technical skills in any industry. Their ability to go on site and the demand for their services has forced other, less qualified ''trainers'' into the field. It is important to recognize the systemic nature of illness and injury. Cumulative trauma disorders may result from other causes (e.g., diabetes, tumors, lupus). Many unionized skilled trades employees profess to be experts in the area of ergonomics. It should be noted that the ''tool'' or ''machine'' may not be the cause or source of the trauma to begin with. Only an experienced clinician can identify and prevent common job-related injuries. The registered occupational therapist (OTR) is the most qualified by nature of education and ability to break each job down to its essential job tasks. The OTR will assess the number of repetitions, the weight lifted, motions, postures, and the impact on soft tissues in relation to the body and workload.

Medical and orthopedic, neurological, and physiological conditions are evaluated by the OTR in relation to the worker and the tasks that they perform. The OTR is qualified to categorize jobs as having light, medium, or heavy exertion levels. The U.S. Department of Labor *Dictionary of Occupational Titles—Physical Demand Characteristics of Work* assists the OTR in ranking the work level. The demands of any job classification may also include environmental exposures such as fumes, vibration, and noise levels. The frequency of pushing, pulling, turning, grasping, pinching, crawling, climbing, twisting, standing, sitting, walking, the maximum distance moved, and force are identified. The weight of objects and material handling is graded as never, rare, occasional, frequent, or continual, and the height and shape of the object is determined. The total day is broken down to its smallest task in ranges and degrees. The job requirements and subjective and objective symptoms of the patient/employee is evaluated by the OTR in a functional capacity evaluation (FCE). Care should be given by management in ordering a preplacement evaluation, FCE, job-site analysis, or ergonomic job-site assessment. In any of these capacities, only a qualified, experienced clinician should be performing and documenting recommendations to the patient/employee and employer. A case-by-case evaluation of worker requirements along with the physical demands and essential job functions is performed.

This process is crucial to understanding why some companies demonstrate commitment and process change in search of competitive success. The OTR knows what aspects of the work environment may be too expensive to modify (perhaps due to a new model changeover at a plant within the year). A line change that is a multimillion dollar investment and will occur in a few months may delay the ergonomic change. Workers symptoms from "use" may be decreased by posture or positioning the body part during line changeovers. A splint or other adaptation may be suggested instead. Tradeoffs must be made by those who design and implement change. Ergonomists who are engineers work with occupational therapists to design tools to assist in the patient/employee recovery process or to prevent injuries from occurring. Each procedure reinforces process innovation. The benefits to worker and employer are continuously reviewed. Design engineers assist the OTR and other clinicians in identifying capital budget limits and the rate of improvement in technology. If top management regards the risks associated with changing its process technology as being greater than the product risks, they will—consciously or not—raise the hurdles applied to proposed investments in process improvements. Most managers recognize the best processes for their workers or producers from a business perspective. An experienced, confident, well-prepared occupational therapist and physician will most inevitably successfully fit the worker to the job.

The occupational therapist tests upper body lifting capability and determines the patient/worker maximum physical exertion level, keeping the pathology of the injury or illness factors in check. The engineer has knowledge of the manufacturing system design, while the therapist/ergonomist has the health science added component. In 1990, a Board of Certification in Professional Ergonomics (BCPE) was established in Bellingham, Washington. BCPE offers two certifications: the Certified Professional Ergonomist (CPE) designation and the Certified Ergonomics Associate (CEA) certification. A CPE has a master's degree in ergonomics or human factors engineering, has at least 4 years of experience in worksite or workstation designs, and has passed the BCPE's three-part exam. The CEA certification requires a bachelor's degree in a related field, 200 hours of formal contact in ergonomics training, 2 years of experience in the field, and the successful completion of a two-part exam.

Employers are interested in preventing injuries and allowing injured workers and an aging workforce to continue working without reinjury or injury. Combined clinical expertise with ergonomic design of the workplace is an invention that companies can afford and for safety reasons have asked occupational therapists to come on-site to describe the workers' job duties and functional capability in relation to the workers diagnosis.

The occupational therapist understands the techniques utilized to control pain (pain management) as well as education options for patient/worker instruction to live within the pain tolerance level in order to perform in the workplace. The methods of job site analysis with job description and job duties, biomechanical and energy expenditure models, and the NIOSH guide to estimate the demands of work tasks on the upper extremity and back are collectively utilized by on-site occupational and environmental medicine physicians, registered occupational therapists, and ergonomists.

APPENDIX 1: VIDEOTAPED ERGONOMIC JOB ANALYSIS REPORT

Company: Automobile Company

Address:

Date of Service: October 31, 1997

Therapists: Eleanor Stewart, OTR, and Jane DeHart, MA, OTR,
Registered Occupational Therapists

Introduction

The purpose of the occupational therapy ergonomic consultation was to provide an analysis of three worksites: 1) Robogate Metal Shop (456), 2) Stock Chaser Leader (R&D Dock 7), and 3) the Wood Mill (Building 16). The occupational therapy analysis was performed to determine the necessity for ergonomic modifications in order to minimize sprains, strains, and other soft tissue injuries. This project was completed in cooperation with UAW–Automobile Company and the on-site Department of Occupational Health and Industrial Rehabilitation.

The process involved analyzing the three worksites in conjunction with observing employees perform their job duties to include an assessment of the tools, equipment, machinery, and physical layout of the work areas. Most importantly, the workers' body mechanics and the physical demands required to accomplish each task were also evaluated. The following information describes the three jobs in relation to the workers' posture and interaction with the environment. Additionally, potential solutions are given to reduce risk factors contributing to sprains, strains, and other soft tissue injuries.

Job Description: Robogate Job-Metal Shop 456

Problem 1

Observed worker with both arms fully extended overhead to remove weld gun (2Y-6118) weighing 35 pounds, which was suspended from counterbalancer and attached to a cable. The overall weight of cable was 42.2 pounds and length was

168 inches. This task involved stretching arms above the head to grasp the weld gun for controlling, guiding, and pulling the cable in to position for the "weld shoot."

Solution 1

Minimize the distance between the worker and the object lifted. The worker's body should be as close as possible to the object to be lifted. Keep lifts between waist and shoulder height whenever possible. In order to reduce neck and shoulder stress, the weld gun and cable should be handled at the point closest to the worker, maximizing the length of gun handle. The gun is air-controlled to minimize force. In addition, eliminate extreme and overhead reaching by relocating the counterbalancer to mid-chest height rather than overhead.

Problem 2

Observed the worker bending at the waist while twisting his trunk and back to place the welding gun toward left side of the front window using weld gun (HL62-2).

Solution 2

Use proper body mechanics during lifting, bending, and material-handing tasks. The recommended technique is to tighten abdominal muscles and bend at the hips/knees rather than at the waist. Tightening the abdominal muscles and bending the hips and knees while lifting is recommended rather than bending at the waist. This keeps the back in correct alignment. The back has three natural curves: cervical (neck), thoracic (middle back), and lumbar (lower back). When all three curves are balanced, the ear, shoulder, and hips should line up straight. Proper body mechanics or "good posture," in fact, means keeping the back curves in balance, reducing back strain and injury.

Problem 3

Overhead pulley weld gun (4Y4181D) weighs 30 pounds, the cable length is 168 inches, and weight is 45.2 pounds. This motion can cause significant shoulder and arm fatigue as workers are required to reach overhead for prolonged periods of time. The greater the reach, the shorter the endurance time, causing muscle fatigue. The position of the head and neck is usually determined by the visual requirement overhead. Often when performing precise motions the workers are required to lean forward to obtain good vision of the work area. Also, forces are placed on the cervical extensor muscles while forces acting on the spine and the levator scapula muscle were observed. Adjustment of the height of the weld gun and cable is necessary to accommodate the worker standing in erect posture. The weld gun cable could be retractable to reorient the work to eye level as previously noted regarding relocating counterbalancer. Again, the retractable arm would allow the weld gun to be at eye level based on the actual height of the individual working at the work station.

Solution 3

Install a "skillet" (12 ft × 18 ft) that rotates to reduce overhead work and allow proper positioning of head, neck, shoulders, and hands. A foot pedal can be utilized to rotate and adjust the frame of the car. Build the car from the bottom up to eliminate overhead reach and awkward postures.

Problem 4

Lifting and handling the weld gun involved using the shoulder, elbow, forearm, wrist, and hand to perform certain motions. The physical stresses result from pressure placed on the soft tissues such as muscles, tendons, nerves, and blood vessels, especially involving the thumb, four digits, palm, or surface of hand and forearm as observed during this task.

Solution 4

Keep elbow in close alignment to the body or elbows in at sides. The wrists should be kept in neutral or 0–30 degrees of flexion when possible. This motion is best depicted when an individual opens a refrigerator door, in which a neutral posture is used that is the correct motion for the wrist.

Problem 5

Forces were produced on the muscle, myofascial sheath, tendons, ligaments, and bones when guiding, lifting, holding, and carrying the weld gun against the force of gravity and its own reacting forces to weld.

Solution 5

The balancing of the weld gun at mid-chest, and installing an overhead chain and sprocket conveyor system will assist in decreasing forces produced in this job as this system will move weld gun to the point of use. The workers should perform a full hand prehension, pulling the object so that the palm forms one of the gripping surfaces without flexing the wrist or pulling the wrist in an ulnar deviation pattern. Objects (e.g., tools and equipment) that are held in the hand should be retrofit with protective materials for shock and vibration absorbance and should have well-rounded corners. Wrist posture can be controlled by using a pistol-grip handle, by tilting the weld gun, and by lowering the cable to facilitate an extended or neutral wrist posture.

Problem 6

Workers were observed climbing over the car frame in the area of the motor compartment approximately 2 feet 6 inches high. The gentleman displayed difficulty in balancing while climbing over the frame, requiring hip and knee flexion in excess of 90 degrees for going in and out and stepping onto the platform.

Solution 6

The physical demands of climbing in and out of the car frame can be eliminated by placing the car on the ''skillet'' and rotating it to eliminate the need to climb in and out. This will also eliminate any strain to patient's knees, hips, and lower vertebrae.

Problem 7

Observed worker moving from standing, stooping, kneeling, crouching, twisting, bending, and sitting in order to perform the ''rear wheel-well'' weld. The worker must sit underneath the car frame holding the weld gun in place. A second person assists this person in a crouched position. The elbow is in approximately 150 degrees of flexion and the weld gun is rotated overhead for the welding task. The worker must position himself underneath the wheel frame, which is 38¼ inches high and 31½ inches wide, and fully rotate his body to weld the opposite end in this confined space.

Solution 7

Good body mechanics is required when going from the full body range of motion into the sitting position. It is recommended that the body not lean toward the right or left side: good posture includes upright posture, shoulders relaxed, objects close to the body, with the normal curves of the spine maintained. Head and shoulders should also be upright and elbows and knees bent; the person should not slouch. It is important to maintain good strength and mobility of the back, arms, and legs to prevent back injuries. On-the-job stretching exercises are recommended as listed below:

1. Standing back bends: with feet spread apart and your hands on your hips, slowly bend backwards. Hold for several seconds then return; repeat 3–5 times.
2. Side bends: stand with feet shoulder width apart, slowly bend to the side by reaching downward with one hand and keeping the other hand on your hip. Alternate 5 times on each side.
3. Hip flexor stretch: while holding onto a stable object for balance, reach behind and grab leg around the ankle while maintaining a straight posture; pull up on the leg slowly to stretch the hip muscles and hold for 7 seconds.
4. Standing gluteal stretch: while standing on a level surface, bring one knee up towards chest, grab it with one or both hands, and slowly pull

towards chest until you feel a good stretch in the buttocks area. Hold for 7 seconds.

5. Groin hamstring stretch: bend one knee and lean to the side, keeping the other leg extended and stretched. Push down gently in the direction of the bent knee and hold for 7 seconds. Repeat for the other leg.
6. Back rotations: slowly rotate the body to each side 5–10 times.
7. Toe touches: slowly extend arms and reach down toward toes without bending knees. Repeat 2–5 times.
8. Isometric neck exercises: clasp hands behind head and, using only mild force, push head into hands and prevent movement of the head by resisting with the hands. Maintain the force for 7 seconds, and slowly release at the end. Repeat 2–5 times.

Safe lifting techniques are as follows

1. Move as close as possible to the object and keep feet at a comfortable distance for adequate support while maintaining a neutral inward curve of lower back.
2. Bend at hips and knees while keeping head and shoulders upright and pick up object keeping it as close to you as possible; *do not twist at the waist.*
3. To complete the lift, turn body as a whole unit by repositioning feet.
4. Lower objects by applying the same principles of proper body mechanics as above.

Problem 8

Workers must maintain a static posture while reaching over the top of the car frame to hold the cable, which is 96–168 inches in length. Reaching from outside the car frame over to hold a portion of the cable 24 inches from the shoulder up to the clamp is the measured reach distance. The actual straddle of the width of the car frame is $4^1/_2$ inches. Also, the employee must place his foot on 12-gauge sheet metal while standing on tiptoes to weld.

Solution 8

Avoid prolonged or static positions of the body with frequent postural changes and the avoidance of repetitive activities. Workers should wear shoes that have arch support. Also, a platform or flat surface could be used so that there is a full support for the foot. Stretching and strengthening exercises for the ankle include dorsal and plantar flexion or pointing the toe toward the floor and toward the ceiling as well as ankle circles counterclockwise and clockwise, 5–10 times during exercises.

Job Description: On-Site Job Ergonomic Analysis—Stock Chaser Leader in the R&D Dock 7

Problem 1

Worker was bending back forward and twisting at the waist to unload a pellet (50 inches high and 42.5 inches wide), consisting of office supplies, e.g., 12 boxes of copy paper at 52 pounds per case, laser cartridge, and storage file boxes, etc. The gentleman indicated that each ''stock chaser'' is required to unload two to three pallets daily.

Solution 1

The employee could benefit from body mechanics education and safety training to reduce the risk of back injury. Again, keep the back straight, bend with hips and knees, and keep load close to body. Additionally, a height-adjustable pallet with carousel-turning device to rotate heave pallets with ease is needed to prevent frequent bending and twisting at waist.

Problem 2

The worker was observed reaching with arms fully extended from floor of pallet to overhead while unloading items from the pallet. Measurement of reach distance across the pallet to access packages was 42 inches away from this gentleman's body. Proper positioning, e.g., moving closer to the object, decreases reach distance to 21 inches, or in half.

Solution 2

The employee should avoid extreme and unnecessary reaching, using safety-awareness techniques of keeping objects close to the body, and stand directly in front of the object to be reached.

Problem 3

Observed worker lifting and material handling packages of variable weight ranging from 1 to 52 pounds. He unloaded eight objects from the top of stack on pallet to gain access to cases of paper. These packages were transferred from the pallet to one flatbed for sorting according to items by departments (reroute slip). Next the packages were removed (lifted and handled) from the first flatbed and transferred to a second flatbed for transporting supplies to the departments.

Solution 3

The employee should reduce excessive lifting and material handling via transporting and transferring from the pallet to one flatbed rather than using two trucks. The height-adjusting pallet and an adjustable-height flatbed (used for transporting) will enable the worker to decrease movements involving bending, twisting, and additional reaching and lifting.

Job Description: Model Maker (Woodmill Department, Bldg 16)

Problem 1

The worker was noted flexing head/neck and bending back while gaining access to area containing 80-inch-high wood storage bins, 23 bins across. Model makers are required to pull and cut all lumber according to specifications.

Solution 1

The employee should use correct bending techniques for material handling and lifting, flexing hips/knees versus using back musculature to prevent back injuries.

Problem 2

The worker pushed/pulled a sheet of ''perfect plank'' 4 ft. × 8 ft. (3/4 inch thickness with weight approximately 45 pounds) from storage area and placed on a panel rack cart (without brakes). The sheet of wood was transferred to the saw table a distance of 19 feet away. Other postures observed during transfer involving pivoting, twisting, and turning.

Solution 2

Redesign the wood storage area and install a hoist for lifting and transporting sheets from bin to table saw. This will eliminate bending, twisting, turning, and material handling to transport sheets.

Problem 3

The workers lifted and handled ''perfect plank'' and indicated that lumber varied in weight from 25 to 45 pounds depending on thickness of wood (3/4, 1/2, 3/8, or 1/4 inch). The lifting also included maneuvering and positioning load to be transferred from storage to saw and vice versa.

Solution 3

Refer to Solution 2 to eliminate lifting and handling of ''perfect plank'' and other lumber with installation of hoist.

Problem 4

The worker was observed reaching from ankle to overhead levels with arms fully extended to handle plywood sheet from 1 of 23 bins. Worker stated that two or three men may be available to assist with this task.

Solution 4

Employees should utilize safety techniques for all phases of their job duties, incorporating proper body mechanics and upper extremity positioning while performing reaching and grasping movements.

In conclusion, based on the outcome of the worksite analysis, ergonomic modifications and provisions for body mechanics training are suggested to minimize strain and sprains.

Thank you for the opportunity to be of assistance.
Sincerely,

Jane DeHart, MA, OTR Eleanor Stewart, MSA, OTR

7

Skeletal and Soft-Tissue Injuries and Cumulative Trauma Treatment

An injury/illness is considered work related if the patient reports that the injury occurred while at work. The spectrum of work-related injury/illness includes visits to the emergency departments of nearby medical outpatient ambulatory care facilities and hospitals.

Reports described by job placement and case management give in detail the activities taken by the company during the period since the last injury/illness report, including information on the number of employees seen in the emergency department and the number of those treated for cumulative trauma disease. The control measures implemented to decrease the development of cumulative trauma disease and work practice controls, engineering plans, and training are furnished to local joint safety and health committees.

In a study conducted by Williams et al. (1997) at the West Virginia University Department of Emergency Medicine, involving emergency department patients seen over a one year period, work-related injuries accounted for 1,539 of 12,321, or 12.5% of all injuries. The mean age of the patient injured on the job was 33.8 years compared with a mean age of 27.7 years for all injured patients. Males accounted for 67% of the work-related injury visits compared with 57% of all injury visits. The most common mechanisms of work-related injuries were overexertion, cut or pierced by sharp implements, falls, struck by object, and transportation-related injuries. Sprains and strains were the most common type of injury sustained, followed by wounds to upper limbs, contusions, and fractures. Known medical charges incurred by patients injured at work were as high as

$62,622. The average charge for those treated and released was $273; the average charge for those who required hospitalization was $10,910.

Falls occur usually when workers use tools or during handling and involve ladders, scaffolds, or elevated work surfaces less than or equal to 10 feet according to another study conducted by Cattledge et al. (1966) using the West Virginia workers' compensation and supplemental injury records to assess nonfatal occupational falls from elevated work surfaces in the construction industry. Approximately 63% of the 182 claimants had received some type of fall-protection training.

These studies and others indicate that injuries sustained at work have a likelihood of affecting the musculoskeletal system (i.e., sprains, strains, wounds to upper limbs, fractures). The emergency department encounters a work-related injury/illness patient, stabilizes the patient, or dictates the patient assessment, and contacts the occupational health and industrial rehabilitation team to follow up with care on-site at the workplace. The computerized linkage between the emergency department and the on-site facility allows for prompt communication regarding the patient's medical status. Telephone consultation from a surgeon with the on-site occupational medicine physician and necessary postdischarge planning office visits are available. Clinical pathways limit variation in treating workers' compensation patients and reduce customer employers' costs. Patients are able to return to their worksite with physical therapy and/or occupational therapy offered to them. The patient is able to receive therapy with other workers who are familiar with the worksite and are able to arrive at work fully dressed and alert, unlike therapy at a hospital or outpatient traditional therapy clinic, where patients do not work for the same company, may be dressed in hospital gowns, and may be retired or quite ill. All of the results of specialized tests (e.g., computerized tomography scan, magnetic resonance imaging, myelogram, bone scan) are received by the on-site physician and therapist. The benefits include bringing employees back to work sooner and improved company commitment to employee well-being.

Therapy increases the opportunity for creativity on the part of supervisors in developing work assignments. Coworker friendship increases the likelihood that patients/employees will follow through with their recovery process. To the degree that employers are faced with liability for the care of workers who become disabled for long periods of time or who may become permanently disabled, the access to necessary rehabilitation services needs to be provided conveniently and promptly. Reasonable accommodation includes reassignment to a vacant position, adaptation of equipment, and job restructuring (e.g., part-time and modified work schedules). Flexible scheduling and time off for medical appointments within reason give the patient an opportunity to perform the essential job functions for which he or she is qualified during the course of rehabilitation. Supervisors and managers play a critical role in the injured/ill employees' efforts to return to work. The supervisor is usually the first to know if an employee has sustained the injury at work (by filling out sections of the accident report). The supervisor is the key person in defining the functional conditions under which the work occurs and whether or not the restricted worker can be accommodated. A safe return to work in the department, where the worker must function at 100%, is the goal of the industrial rehabilitation physical therapist and occupational therapist. The on-site model reassures supervisors of the worker's capabilities. The supervisor's attitude is based on accurate information from medical personnel familiar with the work and the costs of work disruption. A patient's work restrictions are known to the supervisor and there is no distortion of information. For example, pain is subjective and a worker with back pain knows his or her limits in accomplishing work duties safely and without pain. Rehabilitation provides clear work expectations following an initial assessment.

A document measure ensures that the right information about the patient's/worker's capabilities to perform the job are identified for reintegration into the workplace. Therapists working on site in rehabilitation may use evaluation forms or follow their own evaluation format. They may use a variety of patient exercise sheets or protocols or set up individual exercise programs dependent on patient need. Industrial rehabilitation values occupation.

The evaluation process in detail and the progress of work-related conditions most commonly seen on site are included in this chapter. Selected evaluation procedures are introduced and application to specific cases are made. The examples are noninclusive, not all-inclusive. Any clinician will administer only evaluation testing procedures that are indicated for a particular diagnosis. Overtesting is not cost-effective in today's marketplace. Measuring outcome is an important part of providing information to support decision making for workers' compensation care and return to work. Figure 1 is an example of an industrial rehabilitation physician referral form.

The following examples illustrate an industrial rehabilitation initial evaluation used in physical and occupational therapy. Samples are used for format purpose only, not for content.

DEPARTMENT OF OCCUPATIONAL HEALTH
AND INDUSTRIAL REHABILITATION

Patient's Name: ____________ Medical Record Number: Date:

Diagnosis:

Precautions:

Frequency_ per week X_ no. weeks (or number of treatments)

PHYSICAL THERAPY

- ☐ EVALUATE AND TREAT
  - ☐ Exercise
  - ☐ Back program
  - ☐ Myofascial release and trigger point release
  - Modalities:
    - ☐ Ultrasound
    - ☐ Moist heat
    - ☐ Electrical muscle stimulation
    - ☐ Iontophoresis
    - ☐ Phonophoresis
    - ☐ Paraffin bath
    - ☐ Hot/cold therapies
    - ☐ TENS
    - ☐ Whirlpool
  - ☐ Lumbar MedX treatment
  - ☐ Massage
  - ☐ Jobst Pump
  - ☐ Cervical and Lumbar Traction
- ☐ OTHER________

  ____

  ____

OCCUPATIONAL THERAPY

- ☐ EVALUATE AND TREAT
  - ☐ Hand therapy
  - ☐ Range of motion
  - ☐ Strengthening
  - ☐ Coordination
  - ☐ Edema control
  - ☐ Desensitization
  - ☐ Activities of daily living
  - ☐ Customize hand splint/sling
  - RETURN TO WORK CENTER
  - ☐ Functional capacity evaluation
  - ☐ Work conditioning
  - ☐ Job site analysis
- ☐ Job Coaching
  - ☐ Ergonomics adaptation and education
  - ☐ Prevention, proper posturing
  - ☐ Back program

Contraindications

Remarks________________________Physician's Signature

For On-Site Team Use Only:

Patient Address

Street____________ City______ Zip______ Job Title and Shift

Phone (w)___________ (h)____________ Work Location

Insurance________ Case Manager_______ Date of Injury

FIGURE 1 Physician prescription for physical therapy and/or occupational therapy.

EXAMPLE 1: ON-SITE INDUSTRIAL REHABILITATION PHYSICAL THERAPY: INITIAL EVALUATION

Database:

Patient Name: ______________ Medical Record Number: _______

Date: _______

Diagnosis: Herniated nucleus pulposus Onset Date: 5/1/99

Ins: Workers' Compensation

Treatment Length: Anticipated 4 weeks

Physician: Dr. Kim

Location: Automotive Industry Plant/Transmission Assembly

History of illness: The patient reports a history of back pain which significantly increased in mid-May while lifting a heavy welding gun. He is on a work restriction of ''no lifting greater than 25 pounds'' and for a 5-hour work day, 5 days/week in transitional work, light-duty conditioning at the plant.

Pertinent Medical History: Tonsillectomy age 5, right Colles' fracture status/post 10 years ago while skiing downhill in Aspen, Colorado.

Surgical History: None reported.

Medications: ''Motrin 200 mg. as needed for pain'' per patient. He states that he has 30 pills left and will return to his doctor during the next 4 weeks for a recheck. He also takes a multiple vitamin daily and an occasional aspirin per patient report.

Subjective

The patient reports difficulty sleeping. He reports he limps after walking short distances, less than 100 feet, pain down the inside of the right thigh, occasionally down the outside of the thigh to the outside of his foot. He reports the pain is a burning pain with sometimes significant shooting pain. Lying down on his back decreases his pain as does wearing a back brace. Ice relieves some stiffness and pain in the back. Sitting in a car or driving irritates the back. He reports a ''bruised feeling over the back and right hip.'' Patient rates right leg pain at a level 4 on a scale of 0 to 10, with 10 being the highest. He states his back pain is at a level 5 ''softball size'' in the middle of his back. Patient is working in light duty in an occupational therapy transitional work reconditioning program. He tolerates a 5-hour workday in this capacity 5 days/week.

Objective

Range of motion: Lumbar flexion 25% of normal limits, extension 25% of normal limits. Right side bending 50%. Left side bending 25% with some complaints of radiating pain. Right rotation 25%, left rotation 25%.

Strength: Hip flexion left 4+/5, right 4/5. Hip extension left 4/5, right 3+/5. Abduction left 4+/5, right 4/5. Adduction right 4/5, left 4/5. Knee flexion, extension 4/5 bilaterally. Right dorsiflexion 3+/5. Low back strength 3/5 within range of motion.

Sensation: Decreased to light touch and pinprick right lower extremity in the L4–L5 dermatomes. Special tests: positive straight leg raise on the right; positive slump test on the right.

Palpation: Complaints of pain over the lumbar paraspinals and right piriformis. Distraction decreased complaints of pain and numbness in the right lower extremity.

Ambulation: Ambulation with antalgic gait on the right, toe drag on the right during initial entry into the department.

Reflexes: Slight decreased patella tendon reflex and achilles tendon reflex response on the right.

Assessment

The patient presents with acute herniated disc along L4–L5, possibly L5–S1.

Problems	Measurable rehab goals
1. Decreased lumbar range of motion.	1. Increase lumbar range of motion within functional limits in 6 visits.
2. Decreased strength lower extremities and low back.	2. Improve strength to 5/5 bilateral lower extremities and low back within 8 visits.
3. Pain in bilateral lower extremities.	3. Eliminate complaints of pain in lower extremities within 8 visits.
4. Numbness, decreased sensation right lower extremity.	4. Eliminate complaints of right lower extremity numbness within 8 visits.
5. Difficulty ambulating.	5. Ambulate with normal gait pattern within 8 visits.
6. Difficulty with activities of daily living (ADLs) (i.e., getting in and out of the car and tying shoes).	6. Independence with ADLs within 6 visits.
	7. Independence with home exercise program within 1 visit.
	8. Return to same job in transmission assembly in 30 days.

Treatment Plan

Hot pack, ultrasound, pelvic traction, lumbar and lower extremity strengthening exercises, posture and body mechanic instruction.

Frequency and Duration

Two to three times a week for 4 weeks, or 8 visits.

Rehabilitation Potential

Good.

Recommended Physician Recheck Date:

Every 2 weeks.

Registered Physical Therapist: ______________

EXAMPLE 2: ON-SITE INDUSTRIAL REHABILITATION OCCUPATIONAL THERAPY—EVALUATION

Database

Patient Name: ___________ Medical Record Number: ___________

Date: _______

Diagnosis: SP Tenosynovectomy due to acute tenosynovitis of the extensor tendons right hand

Onset of Injury Date: 10/25/99 Date of Birth: 12/22/56

Insurance: Workers' Compensation

Prescription Length: 2 × Wk 4 Wks (8 visits) Physician: Dr. Kim

Location: Automotive Industry, Plant Location

Subjective

The patient reported that she repairs transmissions and uses power tools on her job. She reported that she had continual episodes of swelling and was treated initially in the plant's main medical unit with whirlpool and ice. Apparently she was also treated with cortisone injections to relieve the pain from the synovitis. She had surgery by Dr. Kim on 10/25/99. She reports pain at a level 8 or 9 with motion and a level 4 at rest on a scale of 1 to 10, with 10 being the highest pain. She is currently not taking any medicine; however, she reports that medication upsets her stomach. She reports that she is at work 2 hours a day with a restriction of "no use of right hand."

Past medical history includes left ganglion, which still persists, left middle finger amputation distal to the interphalangeal joint at age 9, cesarean section during the delivery of her daughter, two laparoscopies on the right ovary, and appendix removed in 1998 due to right ovarian cyst and possible infection to the appendix.

Objective

Patient is right hand dominant. The involved extremity is the right hand. She has edema throughout the dorsal surface of the hand proximal to the metacarpal phalangeal joints. She has numbness of the thumb tip on the volar surface, and over the scar she has hyperesthesia. Scar adhesions are present at the site of surgery on the dorsal surface, radial aspect of hand. The scar is 4 centimeters in length. Fine motor coordination using the 9-hole peg test shows right-hand speed to be 4 minutes and left-hand speed 1 minute 45 seconds to place and remove the pegs from the form board.

Active range-of-motion measurements are as follows for right dominant hand; the left hand is within normal limits.

Wrist extension	0/43 degrees
Wrist flexion	0/63 degrees
Radial deviation	12 degrees
Forearm supination	0/78 degrees
Pronation	0/78 degrees

	Metacarpal phalangeal	Proximal interphalangeal	Distal interphalangeal
Index finger	0/66 degrees	0/105 degrees	0/47 degrees
Middle finger	0/80 degrees	WNL	0/52 degrees
Ring finger	0/78 degrees	WNL	0/61 degrees
Little finger	WNL	WNL	0/66 degrees

Thumb CMC and MP joints are WNL. Thumb IP joint flexion was 7 degrees. Thumb volar abduction was 0/67 degrees. Radial abduction was 0/63 degrees. Opposition is within normal limits.

Activities of Daily Living: patient is independent with all dressing, bathing, and hygiene.

Volumetric measurement: right hand, 549 ML milliliters; left hand, 492 ML milliliters.

Sensation: hyperesthesia over the scar from surgery.

The patient's grip strength taken with the dynamometer on the second setting was 19 pounds on the right dominant affected and 58 pounds on the left. Patient's pain level on a scale of 1 to 10, with 10 being the highest, was a level 8 upon motion per patient report, thus limiting the evaluation. The pain remained at an 8 postevaluation.

Right lateral pinch prehension strength was 16 pounds; left was 17.5 pounds. Right 3-jaw pinch strength was 11 pounds; the left was within normal limits at 27 pounds.

Assessment

The patient has had a tenosynovectomy to prevent rupture of the extensor tendons. She has been referred for occupational therapy, range of motion, strengthening, and coordination.

Problems:	Measurable rehab goals
1. Decreased active wrist flexion and extension and radial deviation	1. Increase wrist extension to 70 degrees, flexion to 80 degrees, and radial deviation to 20 degrees actively in 8 visits
2. Decreased forearm supination and pronation	2. Increase forearm supination and pronation to 90 degrees actively in 8 visits
3. Decreased digit 2 through 3 MP joint flexion; decreased index finger PIP joint flexion; decreased digit 2 through 5 DIP joint flexion	3. Increase MP joint flexion to 90 degrees, index finger PIP joint to 110 degrees, and DIP joint flexion to 90 degrees actively
4. Decreased thumb IP joint flexion; decreased volar abduction slightly	4. Increase thumb IP joint flexion to 80 degrees; increase volar abduction to 70 degrees within 4 visits
5. Decreased right-hand lateral pinch prehension, 3-jaw pinch prehension, and grip strength, and decreased fine motor coordination	5. Increase right-hand lateral pinch prehension strength to 25 pounds, increase 3-jaw pinch prehension strength to 20 pounds, increase grip strength to 60 pounds, and increase fine motor coordination within 8 visits
6. Edema, scar adhesions, hyperesthesia, and some numbness in the tip of the thumb	6. Decrease edema through retrograde message, positioning, and hand-pumping; decrease hyperesthesia through desensitization techniques; decrease scar adhesions through cross-friction massage and controlled vibration massage; provide the patient with a home exercise program
7. Patient is not working	7. Return patient to the job of repairing transmissions in a work-reconditioning program; building for 2 hours/day to working toward a full 8-hour shift, following the 4-week hand therapy program (within 30 days)

Recommend occupational therapy for 8 visits, 2–3 times per week, and follow-up with Dr. Kim in 2 weeks.
Prognosis is good. Thank you, Dr. Kim, for this referral.

Occupational Therapist: ______________

Suggestions for providing O.T. workers' compensation care follow.

I. Data Gathering

Initial information will be gathered from the patient, the medical record, patient's family (if indicated), case manager, workers' compensation representative, job placement coordinator, and the referring physician. Relevant information will be recorded in the patient's O.T. evaluation and progress reports.

According to the diagnosis and the patient's condition, the following evaluations should be performed:

1. Upper extremity active and passive range of motion
2. Upper extremity muscle strength
3. Upper extremity sensation (protective and discriminative)
4. Upper extremity fine and gross motor coordination
5. Upper extremity edema
6. Activities of daily living (feeding, dressing, bathing, hygiene, and homemaking) and postural stresses (push/pull, grip, lift)
7. Vocational assessment
8. On-site job assessment as indicated
9. Functional capacity evaluation

The following areas are also to be considered during data gathering and evaluation:

1. Injury history (date of occurrence, circumstances of occurrence, date/type of subsequent medical intervention structures involved in condition)
2. Past medical history and medications
3. Personal history (age, work history, sex, home environment, support system, home responsibilities)
4. Vocational history
5. Pain
6. Splint use, protective equipment, ergonomic modification
7. Patient goals and motivation
8. Demonstration of home exercise program

II. Treatment Programs

A. Throughout the patient's treatment program, medical advice regarding precautions and/or physical restrictions should be obtained from the attending physician. The attending physician will be kept apprised of the patient's status.

B. Graded restoration activities, exercises, modalities, and mobilization techniques will be provided to improve the patient's upper extremity range of motion, strength, coordination, and sensation for return to work.

III. Documentation

A. The initial note is to include results of evaluation in areas specified under Section I, as well as assessment of the data, treatment goals, plan, and recommendations.

B. Daily and progress notes will include documentation of patient's actual performance related to areas noted in evaluation and treatment plan. It will include data of reevaluations along with revised goals, plans, and recommendations.

C. Any equipment patient uses in treatment or is provided with for home use will be documented.

D. Any instruction given to family members will be documented.

E. Physical fitness and psychological factors will be documented.

F. Content of interaction with team members will be documented.

G. Anticipated date of recovery for return to work.

I. Hand Rehabilitation Program

The hand rehabilitation program was established on site at an automotive industry plant for employees/patients and their family members. The program provides intensive occupational therapy for those patients with impairment of function due to injury or disease. Occupational therapy plays an important role in the prevention of deformity and permanent disability. At present, these services are offered to diverse businesses including the auto, steel, financial, food, beverage, and service industries, to name a few.

II. Established Professional Practices

This program is based on established occupational therapy methods of practice including splinting, range of motion, strengthening sensory testing, eye-hand coordination, motor control, edema and pain management, tool use, on-site job assessment, job coaching, work reconditioning, and functional capacity evaluation. Clinical guidelines, pathways, and competency-based therapy methods, techniques, and modalities are adhered to.

III. Quality of Treatment

Quality of treatment is assured by supervision and audit through the on-site provider of industrial rehabilitation at the company. A medical advisory board and grand rounds/case management meeting is held with all members of the medical management process involved and working as a team. Quality assurance monitoring is ongoing.

IV. Objectives

A. To prevent or minimize residual, physical and/or psychological disability through:
   1. Splinting and adaptive equipment as indicated.
   2. Programmed functional outcome activities to increase physical and sensory function (i.e., range of motion, coordination, sensory reeducation, muscle strengthening, physical tolerance, etc.).
   3. Work assessment.
   4. Job coaching and ergonomic redesign of a work station.
   5. Worker and company education.

B. To provide the patient with sufficient knowledge to understand his/her disease, how to live with it and continue gainful employment within the limits of his/her capabilities.

C. To enable the patient to gain independence in work job tasks through training and instruction in adaptive techniques or ergonomic modification.

D. To enable the patient to perform work responsibilities through instruction in pain management, posture, work simplification, energy conservation techniques, body mechanics, and prevention.

E. To return the patient to optimum living and/or function.

F. To decrease work restrictions.

G. To transition patient to full duty through work-conditioning programs.

V. Policies and Procedures

A. Policy: All patients/employees must enter the rehabilitation program through a physician's referral and medical supervision.

B. Procedure: Upon referral, evaluation and/or treatment will be completed by a registered, board-certified occupational therapist. Documentation of evaluation, treatment progress, and discharge/follow-up summaries will be maintained and submitted to referring physicians and third-party payers as requested/required.

1. Written documentation includes:
   a. Patient/employee name, diagnosis, referring physician/agency, past medical history and medications.
   b. Diagnosis, date of onset, precautions, associated problems, restrictions.
   c. Initial evaluation.
   d. Program plan, progress, changes, interim evaluations.
   e. Home program, discharge plan, follow-up, family interaction, physician and/or team interaction, education, ergonomic modification.
   f. Third party forms.
   g. Work status.
2. Verbal reporting includes:
   a. Evaluation data, progress.
   b. Education/instruction to the patient/family members, union, job placement, supervisors, human resources, benefits representatives, and case management team.
   c. Phone information on progress/status to insurance carriers, allied health professionals, etc.
   d. Referrals to outside sources (e.g., vocational retraining programs).
3. Each individual evaluation includes the following areas as indicated by disability:
   a. Range of motion
   b. Muscle strength
   c. Muscle tone, postural stresses (i.e., push/pull, grip, and lift)
   d. Motor control/coordination and walking, standing, sitting, bending, reaching, push, pull, crouching, squatting, and lifting in full body range with graded levels of weight (i.e., 0, 1, 2, 3, 5, to 150 pounds); functional capacity evaluation for job duty compatibility
   e. Edema
   f. Sensation: proprioceptive, tactile, stereognosis, two-point discrimination, Semmes-Weinstein monofilaments
   g. Potential for functional use of extremity
   h. Scar adhesions

   i. Psychological factors that might affect program:
      Premorbid social and vocational history
      Motivational level postinjury
      Personal acceptance of disabilities, splinting, or adaptive equipment
   j. Activities of daily living, job task analysis, and onsite job assessment
   k. Patient satisfaction surveys and recommendation to the fitness/wellness center for outcome and prevention programs

   Ongoing evaluation during treatment is an inherent function to determine the effectiveness of treatment and need to upgrade or change the program. The time-limited, measurable outcome is documented.

4. Treatment is determined by results of evaluation and based on realistic functional goals. Treatment may be directed toward:
   a. Decrease of edema
   b. Increase of range of motion
   c. Positioning via education and/or splinting
   d. Strengthening
   e. Sensory reeducation/desensitization
   f. Reduction of scar tissue
   g. Change of dominance, if indicated due to restriction
   h. Increase of coordination, dexterity, force, lift, standing, sitting capability
   i. Independence in self-care
   j. Increase in balance and mobility
   k. Decrease of pain
   l. Increase in duration, weight lift capacity, and repetitive motion to perform job at full capacity
5. Treatment and modalities include:
   a. Edema control, compression garments, massage, and elevation
   b. Range of motion and joint mobilization
   c. Dynamic and static splinting
   d. Muscle and sensory reeducation techniques, myofascial release, and trigger point release
   e. Graded activities and exercises for working through pain and stiffness, blocking exercise, tendon, and nerve glide exercises

f. Skilled activity related to former job skills or specific functions
g. Grip and pinch strengthening, intrinsic and extrinsic stretch, and strengthening
h. Worksite prevention and posturing techniques
i. Prosthetic and orthotic training
j. Desensitizing techniques, scar management
k. Modalities (i.e., heat, phonophoresis, Tens, cold, ultrasound, paraffin, whirlpool, iontophoresis, electrical stimulation)
l. Work conditioning and job coaching

VI. Discontinuation or Discharge

A. Discharge evaluation is administered to determine the effectiveness of the treatment program.

B. Vocational rehabilitation referrals and evaluations are coordinated directly by the physician and therapist with the local office of the State Department of Education, Vocational Rehabilitation Service, if indicated. The third-party payer and case manager also coordinate these efforts.

VII. Documentation

A. The initial note is to include results of evaluation in areas specified in the treatment plan.

B. Progress notes will include documentation of patient's actual performance, results of reevaluations, and noted changes in the patient's condition as related to areas noted in evaluation and treatment plan.

C. Any splints or other ergonomic equipment provided to the patient or used in performance in treatment must be documented.

D. All recommendations given for performance and equipment or home programs must be documented.

VIII. Feasibility

A. Financing: Occupational therapy treatment charges are based on individual treatment rates by units of time, or therapists salary, or a fixed monthly management fee. Additional equipment (such as splints) may be priced at a separate or an additional charge. This program is reimbursed by most third-party payers and company purchasing departments connected to plant or the operational budget.

B. Resources

1. Equipment: Most equipment is available at the on-site rehabilitation clinic.

2. The most recent rehabilitation and therapy techniques and text books serve as a resource for specific protocol treatments as prescribed.

Occupational therapists affect the future delivery of our services (i.e., industrial rehabilitation) with the ability to customize hand splints. No other profession is more relevant to industrial therapy services than the skills of an occupational therapist.

SPLINT EVALUATION*

If the splint is well designed and of optimal value to the worker, the following questions should be answered "yes."

1. Does the splint conform to and maintain the normal transverse and longitudinal arches of the hand and wrist?
2. Does the splint position the wrist in 5–10 degrees ulnar deviation? (This only applies if the worker can tolerate this position.)
3. When the splint is worn for a half hour and used in a job task, is the worker's hand free of persistent redness caused by splint or strap pressure?
4. If metacarpal phalangeal flexion is desired, does the palmar end extend only to the distal or mid-palmar crease? If fourth and fifth metacarpal rotation is desired for power grasp, the splint should extend one-half inch proximal to the distal palmar crease or to the "proximal palmar crease."
5. If metacarpal phalangeal joint protection is desired, does the palmar end extend to the middle of the proximal phalanx?
6. When thumb motion is desired, does the thenar clearance allow for full opposition?
7. Is the splint sturdy enough to provide the desired stability for the wrist when the worker is using his hands?
8. Does the fit of the splint allow for distribution of pressure over the widest area possible, and is it free from nerve compression?
9. When fitted, is the worker free of any pain caused by the splint? (It should not push the wrist into too much dorsiflexion or cause pressure over the ulnar styloid.)
10. Can the worker put on and take off the splint without causing stress to the opposite hand?

---

* Lecture to ergonomic lab at major automobile company headquarters.

Optimal Splint Fit

Suggestions for achieving optimal fit of the workers' splint are listed as follows:

1. Ulnar styloid: For workers with tenosynovitis and/or wrist involvement, pressure over the ulnar styloid can be prevented by padding the styloid area on the splint or on the worker before shaping the splint or by keeping the ulnar styloid free.
2. Swelling: Some workers have swelling. For example, they may swell consistently during the night, or while working. A splint fitted for these workers during the middle of the day needs to be designed in a manner to accommodate the increase in swelling during the 24-hour day.
3. Perspiration: Some splinting materials have a high sweating factor; the use of talcum powder or stockinette liners can increase wearing comfort and prevent skin irritation at night. It may be helpful to show patients how to make their own liners so that they can keep them clean. Perforated splint material is recommended to allow the skin to breathe.
4. The metal stay can be removed from a wrist splint (i.e., freedom splint) to prevent force while performing a job task (such as keyboard entry). Some splints can be molded for proper wrist extension (i.e., plastic stay).

Patient Education

A splint is of value only if it is worn correctly. Worn incorrectly the problems that may occur secondary to the use of a splint may be detrimental to the patient. Educate and instruct the patient on the points listed below in order to allow for proper splint use and fit:

1. The purpose of using the splint at work and for the patient's hand symptoms.
2. Frequency and duration of the splint wearing.
3. What exercises to do in conjunction with the splint use. This is important during the immobilization phase.
4. How to take the splint on and off/donning and doffing.
5. How to determine if the splint is positioned correctly on the hand.
6. How to clean the splint.
7. How to check the skin for pressure areas, redness, or tightness around a bony prominence.

Patient Objectives

The worker should be able to demonstrate the following home program:

1. Tell the therapist the purpose of the splint and wearing schedule.

2. Demonstrate home exercises (which may include passive range of motion, active range of motion, tendon glide, hand pumping, digit opposition, claw hand intrinsic/extrinsic stretch and strengthening, and all thumb motions).
3. Put the splint on and take it off properly without any verbal or nonverbal assistance from the therapist.
4. Determine if the splint is positioned correctly.
5. Explain to the therapist what has to be done to care for the splint.
6. Describe what pressure areas of the skin look like.

OUTLINE OF BACK INJURY PREVENTION WORKSHOP

1. Use a black or white board for demonstration of back physiology to go with the skeleton.
2. Discuss risk factors:
   Diet
   Lack of exercise
3. Discuss methods of force reduction.
4. Torque = Force × Distance: demonstration of how this simple physics equation can have a profound influence on the workload.
5. Practice lumbar stabilization postures.
6. Practice wall squats and how legs can carry the workload.
7. Principles of prevention:
   Twisting
   Overhead
   Lower loads
8. Plan the lift group participation with work stations—squat with the lift.
9. Balance—how to keep good stance during lifting practice session.

HAND EVALUATION PROTOCOL

Diagnostic history:

1. Medication schedule
2. History (onset of disability, onset of disease)
3. Joints involved
4. Type of arthritis noted
5. Duration of morning stiffness
6. Pain—at rest and/or during movement
7. Range of motion/posture/restrictions

Problems identified:

1. Patient's goals
2. Functional capacity evaluation
3. On-site job assessment
4. Hand assessment

Leisure/Work: How injury/illness affects performance in work and leisure

Desire to work
Vocational history
Job title
Job tasks required
Duties unable to complete
Duties able to complete with assistance
Job restrictions (current and past)

Deformities noted: Indicate digit and joint location for

Mallet deformity
Boutonnière deformity
Swan neck deformity
Claw hand
Synovitis
Osteophytes (Bouchard's and Heberden's nodes)
Ganglions
Ulnar drift
Subluxation
Crepitation
Raynaud's phenomenon
Bony resorption
Tendon rupture
Trigger finger
Ulcerations, wounds, amputations, skin color, joint integrity

Major limitations:

Inflammation
Pain and sensation
Intrinsic/Extrinsic tightness
Scar
Wound
Activities of daily living (dressing, bathing, hygiene, ability to drive and work)
Grasp/Prehension patterns

Coordination
Strength
Grip
Pinch
Muscle strength (including intrinsic muscles)
Special tests such as Froment's sign, Finkelstein's test, Phalen's test, and Tinel's sign, among others
Atrophy of muscle
Trigger finger or thumb
Congenital anomalies
Ulnar drift
Dupuytren's contracture
Gamekeeper's thumb and opposition of thumb
Collateral ligaments and volar plate integrity
Aches of the hand

Edema:

Volumeter
Circumference
Fusiform swelling (joint capsule circumference)
Sausage swelling

Assessment

Patient's understanding of disease process
Functional limitations
Nerve compression/entrapment
Need for splinting
Need for adaptive or ergonomic equipment or modification
Home program needs
Summary of findings

Plan

Patient education on:
Joint protection, posture, and mobility
Energy conservation and work conditioning
Active range of motion, stretching, and strengthening exercises, coordination, pain management, and edema control
Splinting plans
Practice with necessary adaptive or ergonomic equipment or modification
Recommendations and prognosis
Frequency and duration of treatment

Goals

1. Increase physical exertion levels, independence in activities of daily living, and work tasks/duties to maximal medical potential
2. Patient education
3. Increase strength and whole body range of motion and functional capacity for return to work
4. Pain management, edema control, desensitization, mobility and work conditioning, among others based on the findings
5. Goals are measurable and time limited

Example from Hand Splinting Lecture

Symptoms	Splint
1. DeQuervain's pain over abductor pollices longus tendon in the 1st dorsal compartment and the extensor pollicis brevis tendons. Pain increases with active or resistive extension and abduction of the thumb.	Immobilization of wrist and the carpal metacarpal and metacarpal joints of the thumb. When fitting the splint, the superficial branch of the radial nerve and the ulnar digital nerve of the thumb must be free of compression. Wrist—15 of extension; carpal metacarpal (MP)—40–50 of palmar abduction; metacarpal phalangeal—5–10 of flexion; the interphalanax (IP) is left free.
2. Extensor tendon injury zones V, VI, VII, T-IV, and T-V.	Wrist in extension 40–45 degrees. The MP and IP joints rest in traction at 0 degrees to prevent extensor lag.
3. Extensor tendon lesion Zones I and II.	Splint position may be volar or dorsal. The distal interphalangeal is immobilized at 0 degrees or slight hyperextension for Zone I injury. Splint immobilization that allows even slight flexion will result in extensor lag. If the proximal interphalangeal (PIP) joint exhibits a posture of slight hyperextension, the proximal joint should be splinted at 30–45 degrees of flexion, while the distal joint is splinted in full extension to advance the extensor mechanism distally by virtue of the central slip.

4. Extensor tendon (injuries in Zones III and IV may result in a boutonnière deformity).	Rupture or attenuation of the extensor tendon over the PIP joint may result if the triangular ligament also is stretched or damaged. Volar displacement and shortening of the lateral bands may also occur. The PIP joint must be splinted at 0 degrees extension. The distal joint is left free to prevent distal joint tightness.

While upper extremity injuries are prevalent at any company, back pain, strain, and sprain are the most common. Occupational back injuries have resulted in prolonged time off of work in the past. Compensation versus noncompensation group studies have indicated significant increases in postoperative workers' compensation visits to specialists and to medical facilities for treatment and procedures. Early intervention in cases of overuse syndrome produces savings. Cases often include patients who have fallen and sustained low back strain. The patient/worker may report decreased treatment efficiency with no relief from pain. There is a need to prevent delayed recovery by performing objective assessment. Health and safety can be a continuing success story. The incident rate for lost workday cases has declined steadily from 4.1 cases per 100 full-time workers in 1990 to 3.3 cases per 100 workers in 1997. In 1997, the rates in manufacturing for days away from work cases and restricted activity only cases were the same (2.4 days away per 100 full-time workers) according to the U.S. Department of Labor. Automobile companies, for example, spend more on the cost of health care than on the price of steel used in their products (i.e., more than $5 billion dollars). The reduction in the number of active full-time employees will cause some decrease in costs. How health care is delivered and the health of the worker as well as worker behavior will shift practice patterns. Age, lifestyle, genetics, union, state and federal regulations, and tort law will drive the results of on-site occupational health and industrial rehabilitation.

A typical case seen on-site is that of a hard-working, 15-year employee of a manufacturing company who slipped on the floor of the plant where he worked and was treated for low back pain. Although the pain was serious, he did not require an emergency room visit. The on-site occupational medicine physician may decide to refer the worker to a specialist for a consultation and magnetic resonance imaging (MRI), but prior to sending the patient to the specialist, physical therapy was ordered for a 3-week time period. The physical therapist educated the patient/worker about techniques and exercises to relieve and prevent back pain. The patient did not need surgery; it was not recommended by the occupational medicine physician because the patient had no leg pain and no objective neurological deficits, especially muscle weakness.

The cooperative approach involving case management of low back pain through the occupational medicine physician with the support of the physical and occupational therapist assisted in the recovery of the patient. The occupational therapist observed the patient performing all his job tasks during an on-site job assessment. Over-the-counter medication (e.g., Tylenol) at normal dosage, Lumbar MedX treatment, myofascial release, 15 minutes of moist heat and ice to relieve spasm, progressive strengthening exercise, and follow-up by the occupational medicine physician within 2 weeks closed the case. The patient had learned to manage his pain, which began at a level 8 on a scale of 0 to 10, with 10 being the greatest pain, and at the end of treatment he reported experiencing pain at a level 3. He had no radiculopathy, nor did he experience any recurring low back pain at any higher level at 6-month follow-up. He was instructed to continue proper exercises in the fitness center to keep in overall physical condition post–physical therapy treatment. The therapist accompanied the patient/worker to the fitness equipment on his first day to eliminate any fear of reinjury on the part of the patient. The patient did not show greater pain during the on-site job analysis. He continually subjectively reported pain at a level 3 with no objective findings or changes during the course of treatment. The patient's motivation for returning back to work in assembly was fair.

The occupational medicine physician monitors all patients with spinal stenosis, low back pain with leg pain (dermatonal variable), limitation with walking, and co-existing with herniated disc, confirmed with MRI or myelogram/CT. The patient may need surgical decompression by a specialist. Back instability with abnormal motion on flexion/extension films and increased pain with activity will be treated with activity restriction, 1–2 days of bed rest, a self-exercise program, medications, self-supplied modalities, and physical therapy with mobilization techniques. If the patient is not progressing, having poor diagnostic studies, facet injection, pain management, or fusion may be recommended by the neurosurgeon or orthopedic surgeon. The physical therapist will objectively measure posture, gait, balance, mobility/range of motion, strength, neurological status, reflexes, sensation, and palpation and perform special tests (i.e., straight leg raising, dual, accessory motions, repetitive movement, joint play, traction/compression, provocation/alleviation). The physical therapist will assess problems of soft tissue, joint, neuromuscular, and deconditioning.

Many approaches to the treatment of low back pain are available (i.e., joint mobilization/manipulation, myofascial release, neural mobilization, Feldenkrais, strain/counterstrain, muscle energy). Robin McKenzie has also developed a protocol that addresses the need to control the recurrent nature of low back pain and teaches the patient to strive for self-reliance by controlling the symptoms through use of corrective postures and/or exercises. McKenzie distinguishes between pain derived from mechanical lesion within the spine and those with pain attributed to inflammatory or other nonmechanical causes. It is the former in which the McKenzie approach is applicable. The fact that the patient is involved in his or her own treatment has enabled many patients to return to work. As with any treatment technique, contraindications (e.g., neoplasms, aneurysms, disease of the bone, central nervous system lesions, worsening neurological signs, instability, or structural anomalies) should be noted. Once pain is controlled and the patient is progressing steadily, the home program of prevention, stretching, and extension exercises may be accompanied by total body exercise including cardiovascular fitness utilizing the treadmill, stair-climbing machine, and exercise bicycle.

The delivery system for managing occupational health includes interaction with case management, providing workers' compensation care, safety and wellness, disability management, integrated information systems, and early return to work strategies. Health care costs, indemnity costs, litigation costs, and reserves (represent funding needed to pay estimated liabilities related to open cases), and redemptions (payments which close or settle a case) are considered when measuring the outcome of returning the worker back to his or her job requirements and duties.

Fitness for duty evaluation can be completed by the occupational medicine physician without the need for independent medical exams or second opinions through provider case management and by knowing how to properly manage injured workers. Appropriate diagnoses and treatment to minimize employee lost work time and to address workers' concerns includes coordinating care with some of the world's finest specialists, which can complement the on-site program and offer credibility to the company contracted customer.

The Henry Ford Health System, which has developed a spinal disease management program with clinical guidelines to manage the full spectrum of back care, is one of five leaders in spinal surgery for outcomes research. The Mayo Clinic, Cleveland Clinic Foundation, University of New Mexico, and Lovelace Health System are part of a consortium collecting data for improving clinical quality, cost-effectiveness, and quality of life. This includes prevention, diagnosis, and the treatment of spinal conditions (Nockels, 1999).

Appendix 1 contains an initial evaluation, daily note, progress note, and discharge summary forms. The occupational therapy services provided on site utilize standard operating procedures as a guideline for on-site industrial rehabilitation. The most common conditions treated include carpal tunnel syndrome, tendon lacerations, DeQuervain's tenosynovitis, cumulative trauma disorders, upper extremity fractures, epicondylitis, nerve dysfunction secondary to injury or disease, tendon repairs, burns, reflex sympathetic dystrophy, rotator cuff/shoulder syndrome and disorders, trigger finger or thumb, tendinitis, ganglion cyst, mallet deformity, and/or others.

APPENDIX 1: ON-SITE INDUSTRIAL REHABILITATION

PHYSICAL THERAPY—INITIAL EVALUATION

Database

Patient Name: ________________________ Date: _______

Medical Record Number: __________ Date of Birth: _______

Diagnosis: ______________ Onset Date of Injury: _______

Insurance: _____________________________

Rx Length: from ______ to ______ Physician: ___________________

Hourly Employee: ____________ Salaried Employee: _______

Current Restrictions: ____________ History of Illness: ____________

Location: ______________ Work History: ____________

Job Title: _________________ Shift: ______________________________

Pertinent Medical History: ____________

Surgical History: ____________________

Medications: _______________________

Subjective

Objective

Assessment

Problem List Measurable Rehab Goals (time limited)

Treatment Plan

Frequency and Duration

Rehabilitation Potential

Recommended Physician Recheck Date

Therapist: ___________________

OCCUPATIONAL THERAPY—INITIAL EVALUATION

Database

Patient Name: ____________________ Date: _______

Medical Record Number: __________ Date of Birth: _______

Diagnosis: ____________________ Onset Date: ____________________

Insurance: _________________________

Rx Length: from ______ to ______ Physician: ____________________

Hourly or Salaried Employee ____ Current Restrictions _______________

History of Illness: _____________ Work Location: _________________

Work History: _____________ Job Title: __________ Shift: ________

Pertinent Medical History: _____________

Surgical History: ____________________

Medications: ________________________

Subjective

Objective

Assessment

Problem List

Time-Limited Measurable Rehab Goals

Treatment Plan:

Frequency and Duration:

Rehabilitation Potential:

Recommended Physician Recheck Date:

Therapist: ____________________

PHYSICAL AND/OR OCCUPATIONAL THERAPY—PROGRESS NOTE

Database

Patient Name: ____________________ Date: _______

Medical Record Number: __________ Date of Birth: _______

Diagnosis: _______________________ Onset Date of Injury: _______

Insurance: _________________________

Prescription Length: from ______ to ______ Physician: ______________

Total Treatments to Date: _____ Location: ____________________

Current Restrictions: ____

Subjective

Objective

Assessment

Treatment Plan/Recommendations for Further Treatment

Home Program

Activities of Daily Living

Frequency and Duration

Rehabilitation Potential

Recommended Physician Recheck Date:

Therapist: __________________

PHYSICAL AND/OR OCCUPATIONAL THERAPY—DAILY NOTE

Patient Name: __________________ Date: _______

Medical Record Number: ________________ Date of Birth: _______

Physician: _______________________ Location: __________________

Diagnosis: _____________________ Onset Date of Injury: _______

Insurance: _______________________

Total Treatments to date: ___________

Subjective

Objective

Assessment

Plan for Changes to Plan with Rationale for the Changes

Anticipated Number of Visits/Frequency and Duration

PHYSICAL AND/OR OCCUPATIONAL THERAPY—DISCHARGE NOTE

Database

Patient Name: ______________ Date: _______

Medical Record Number: _________ Date of Birth: ___________

Diagnosis: __________________ Onset Date: _______

Insurance: ____

Prescription Length: from ____ to ______ Physician: ________________

Total Treatments to Date: _____ Location: ________________________

Current Work Restrictions: _________ Shift: ______________________

Subjective

Objective

Assessment

Treatment Plan/Recommendations for Further Treatment

Home Program

Frequency and Duration

Rehabilitation Potential and Goals Met

Recommended Physician Recheck Date:

Return to Work in Same Job _________ Light Duty Job ___________

Change of Job Duties ___________ Patient Compliant ____________

Therapist: __________________

8

The Prescription for a Healthy Workforce: The Synergistic Whole, or How It All Comes Together

We have all heard over and over again that the future of health care is dependent on the best practices and outcomes. Recently at the Joint Commission on Hospital Accreditation of Healthcare Organizations (JCAHO) survey, the surveyor, Robert Wright, asked me why services in occupational health and industrial rehabilitation were any better than the other local hospitals and for-profit providers in southeastern Michigan and nationally. I responded that I had spent 3 years gathering data on our value. I reported that our aim is to provide workers' compensation injury/illness treatment without the high cost of indemnity (lost work days) and excessive medical procedures in order to be a low-cost provider and increase employee productivity at work.

As evidence of this, measures were developed to track physicians' and therapists' performance. We hold occupational medicine practitioners accountable and report our findings to the employer. Measures included the following studies and/or reports:

- Compare lost work days at companies contracted within southeastern Michigan (Fig. 1)
- Compare restricted work days (banking industry)
- Track medical expenses, reserves, redemptions, indemnity payments (Table 1)

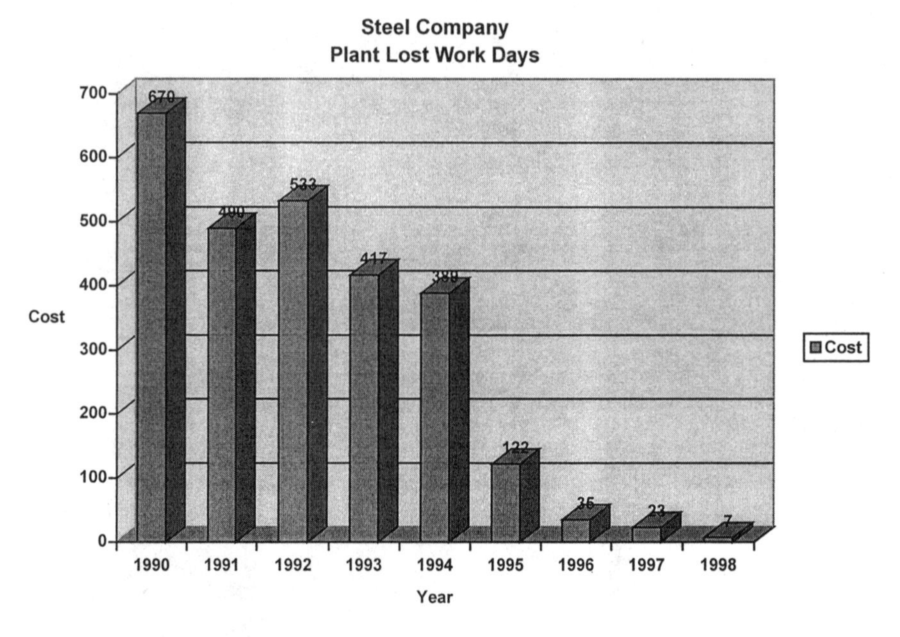

FIGURE 1 Cost of lost work days at a steel company.

Open workers' compensation cases (versus those closed the previous year)
Travel time (workers' compensation mileage paid), number of modalities (heat, cold, phonophoresis, etc.), and charges
Return-to-work rates of patients with back diagnosis (see Appendix 1)
Loss of body fat for those workers exercising in the fitness/wellness center (see Appendix 2)
Growth in on-site occupational health and industrial rehabilitation services (see Appendix 3)

The surveyor wanted to know how we measured success. He was amazed that we had data tracked over 10 years. I told him that we had studied the total number of patients seen in one year with back diagnoses (low back pain, strain, herniated disc, etc.) in industrial physical therapy who utilized the MedX machine to see if patients returned back to work any sooner than the group that had not utilized the machine. Our outcome was a 74.6% return-to-work rate. I then shared the data with the Automobile Company National Benefits Center and the comparison and/or benchmark was made between the other providers utilizing MedX with their "back" population of patients. We were all very similar in percentage of return-to-work rates—approximately ±2 percentage points. The surveyor then

TABLE 1 Department of Occupational Health Worker's Compensation Report Card for a Service Industry Customer

	1 Jan 96	1 Jan 97	31 Dec 98	Variance Fav/(Unfav)	Percent Variance
Open cases	225	152	119	106	47.1%
Reserves	5,058,160	3,822,812	2,879,635	2,178,525	43.1%
			Payments		
		December YTD			
	1996	1997	1998	Cost savings 1997	Cost savings 1998
Gross indemnity	2,160,730	2,019,369	1,651,378	141,361	509,352
Medical expenses	563,861	401,382	363,582	162,479	200,279
Other expenses	411,424	298,092	455,671	113,332	(44,247)
Total payments	3,136,015	2,718,843	2,470,631	417,172	665,384
		Total '97 & '98			$1,083,556
Total savings: reduced reserves and decreased payments					$3,261,081

asked why, if all of the physical therapy providers get the same results, we were any better than the others. I immediately responded that one of our measures showed that we were able to get the same or better results in a shorter time frame with fewer modalities (heat, cold, phonophoresis, etc.). We were able to get these results by utilizing 2.2 modalities when it took our competitors 3.6 modalities to treat the same case. The cost savings in modalities, in fee discounting, and in on-site services at the plant (saving workers' compensation milage expense to off-site facilities) had saved one employer over $7.7 million in the 10 years of tracking. This is evidence-based practice, and it helps us show our employers that we provide workers' compensation care better and for less. It gives us an edge when we compete for new business as we respond to requests for proposal (RFP) to provide on-site or off-site care. The JCAHO surveyor did not stop there with his questions. He wanted to know the next step in making the services provided a cut above everyone else providing occupational health and industrial rehabilitation.

At an automobile assembly plant, to take another example, we have built an 8000-square-foot fitness/rehabilitation center. Patients have been shown a lifetime of "correct exercises" in the fitness center to prevent reinjury. Physical and occupational therapists and athletic trainers also work with patients upon discharge from rehabilitation in the fitness center. The patients often fear reinjury or do not know which equipment to utilize unless the therapist familiarizes the patient with the equipment, posturing, and muscle groups to continue to strengthen. Of course, the use of databases in fitness is also imperative. A prefitness appraisal that records pre-exercise blood pressure, weight, flexibility, body fat composition, and heart rate will help to measure patient benefit and employer cost. The marketplace and health care–purchasing community will reward you for your success rate in controlling costs and, in fact, for measuring quality in terms of productivity and prevention at work year after year. A relationship must exist between the patient, employer, and health care provider. Each must be given respect.

Patient satisfaction can be achieved by implementing a management process for workers' compensation cases. A clinical pathway that is interdisciplinary was developed to outline duration guidelines for the treatment of workers' compensation cases. Defining processes precisely measured by clinical trial can eliminate serious and widespread gaps in time needed to close a workers' compensation case. The occupational health team monitors progress of the workers' compensation job-related injury or illness in grand rounds. The change of effort has improved the quality of care for the specific patient group. Rapid systemized progress has made data available. It is a clear role model for exemplary delivery of workers' compensation care.

Overuse of services or inappropriate services for which a patient does not receive full benefits of the procedure are avoidable. An "activity prescription form'' has been developed for a team of occupational and environmental medicine providers to close workers' compensation cases by documenting appropriate work restrictions, activity, and transitional/limited work duty. Significant improvements can be made with individual patients participating in their own care plan by actively taking part in a work reconditioning program. The individual is able to pace job tasks in accordance with pain level or fatigue factors. The number of complications from returning to work too soon following an injury/illness can be reduced by identifying and treating symptoms early. Expensive complications can be prevented. For example, the use of pain-management programs (medication or acupuncture) or holistic medicine must be evaluated in terms of the patient's health benefit per dollar spent. Variations in the use of treatments may be geographically driven (by a unionized environment), and health practitioner training may be a factor in achieving high quality and low cost.

Our current status is that we continue to collect data. A report card is presented to the direct company contracted customer. We have purchased an automated computerized information system for tracking these outcomes.

Lessons learned include that the pathway has limited variation in treating workers' compensation patients and has reduced customer/employer costs. Patients have returned to work with improved productivity because they return with less lost work time. Employers accept a written report of recommendations regarding ergonomic modifications to the workplace environment.

The JCAHO, the National Committee for Quality Assurance, and ISO-9000 (an international quality standard) have encouraged the use of continuous quality improvement. The ''plan-do-check-act'' (PDCA) cycle has also been utilized to document improvement. Employers may push to control the use of managed care, although the minimum standard of care must have reasonably worked well. Regulation is difficult to change in order to keep up with the latest proven practices. Occupational health and industrial rehabilitation must focus daily on customer-driven outcome.

The department of occupational health and industrial rehabilitation quality manual sets forth the quality system policies and defines compliance with the ISO-9000 Quality System Standard and JCAHO requirements. The quality system applies to services rendered at on-site employer locations. ISO-9000 elements and JCAHO manual sections are documented in a business process flow (see Appendix 4).

The occupational health and industrial rehabilitation model on site is a natural fit for ISO registration because it is an expectation of our customers. Major automotive industries emphasize supplier registration to ISO, and non–health care customers are not familiar with JCAHO standards. ISO provides a framework for documenting (flowcharting) business processes, which increases line staff responsibility and enhances management efficiency. ISO defines processes and redesigns the process for delivering quality services to patients. ISO elements can be built into JCAHO to increase efficiency and show compliance to the standards. The Joint Commission would like to see on-site occupational health and industrial rehabilitation monitor cost, customer satisfaction, and outcome based on ‘‘designing processes,’’ which ISO emphasizes. ISO can be applied to process control and improvement. ISO defines specifications to reduce variation. The department of occupational health will continue to gather data to evaluate the benefits of ISO in terms of outcome, cost, and customer satisfaction. This is a great opportunity to improve organizational performance and process control.

The management responsibility of planning and obtaining business and improving organizational performance is written into the flowchart. Thus, the PDCA cycle is a controlled daily document. Occupational health maintains a documented quality system as a means to ensure that all services conform to specified requirements. Variation is limited and health care practitioners have knowledge of the specifications of the contract, work processes, and objective action plans.

Level One: Quality Manual. The quality manual describes occupational health's department-wide structure and methods for maintaining the quality management system.

Level Two: Quality Systems Procedures (QSP). Documents procedures (flow charts) are used to specify who does what and when it is done.

Level Three: Care Guidelines. Professional practice guidelines and assessments are used by occupational health providers to detail how particular tasks are to be performed where the absence of such instructions would adversely affect care.

Level Four: Records. These are used by occupational health practitioners to provide evidence that the required product or service quality was achieved and that the system has been implemented correctly. Records will include completed forms, assignments, and outcome periods.

The quality system documentation structure simplifies compliance activities and improves organizational performance.

The flowcharts depict the major steps in clinic processes and measure patient outcome. They shows the relationship between the JCAHO manual and ISO-9000 elements.

Occupational health and industrial rehabilitation defines ‘‘customer-supplied product’’ as the ‘‘patient’’ seeking care. The employer provides the patients. The entire occupational health quality system is developed around the concept of continuing care in order to maximize the benefit to the patient. Clinical staff are responsible for identification of all patients and traceability of records. As patients arrive at the respective facilities, they are registered. Assignment of a unique medical record number (MRN) follows, or the patient's Social Security number may be used. Subsequent activities relating to each patient are noted and stored in the patient medical record bearing patient identification. The process control of the patients' treatment is subject to the following:

1. Documented clinical guidelines and policy and procedure for the department of occupational health and industrial rehabilitation.
2. Use of correct tools/equipment in a suitable environment whether the work is patient care or other services (e.g., the use of correct therapy equipment, job site assessments).
3. Compliance with stated care plans and industry, government, or customer-imposed standards, codes, requirements, or procedures (e.g., OSHA, HEDIS, NCQA, JCAHO).
4. Patient care is monitored in real time (e.g., clinical chart review, patient assessment, staff observation).
5. Preventive maintenance (e.g., safety logs, infection control monitoring).

Procedures provide for assessment and monitoring activities that verify that care plans are effective. Records are documented in the patient medical record. Incoming patients are not treated until they have been assessed. The initial assessment provides baseline information in order to plan patient care. Where incoming patients require emergency or urgent care prior to full initial assessment, emergency care may be provided. The patient, when stabilized, will complete the assessment cycle. At all times the patient is identified by MRN or Social Security number.

In-process assessment and monitoring is ongoing throughout the course of treatment. A clinical chart review is provided for each patient. Patients are discharged when it has been determined that they have reached maximum level of function. Evidence that the patient has been properly assessed throughout the care cycle is maintained in the medical record. The medical record indicates assessment results, treatment procedures, modalities, and other appropriate care criteria. The initial evaluation, daily SOAP (subjective, objective, assessment, plan) notes, and the discharge summary defines assessment status throughout care to ensure that appropriate medical care is prescribed. The clinical staff are responsible for ensuring that patients who are not able to return to their regular assigned jobs are identified, evaluated, and continually assessed. The "grand rounds" provides a forum for this assessment. Review of patients on restricted duty is completed per documented procedures and/or work instruction. The patient may:

1. Be rehabilitated to meet the specified work requirements
2. Be allowed to perform regular work duties with or without restriction
3. Be assigned to restricted duty (transitional work or alternate work)
4. Be placed on permanent disability or restriction

Preventive actions may include:

1. The use of appropriate sources of information such as processes and work operations, which affect occupational health quality, audit results, quality records, service reports and customer complaints to detect, analyze, and eliminate potential causes of nonconformities
2. Determining the steps needed to address any problems requiring preventive action
3. Initiating preventive action and applying controls to ensure it is effective
4. Ensuring that relevant information on actions taken including changes to procedures is submitted for management review (process "redesign" for improvement)

Statistical techniques are needed to analyze data collected relating to patient care. The results are reported to the employer. Guidelines and care protocols cut costs by protecting the worker. Our essential priority is to provide exceptional quality and cost-effective care strengthened by excellence in education and research. We work together to improve health and quality of life in the community that we serve. Whether small (45–500 employees), medium (500–4,900 employees), or large (5,000–20,000 employees), occupational health and industrial rehabilitation departments can be tailored to prevent complications and to maximize quality of life. Results utilizing any technique or preferred method of treatment or wellness must be documented to substantiate return on investment.

The Occupational Safety and Health Administration (OSHA) has asked businesses to follow regulations to protect workers from work-related musculoskeletal disorders (WMSD) due to repetitive trauma/stress and overexertion injuries. Ergonomic standards for the manufacturing industries and manual handling operations will need to be established in the workplace. OSHA recognizes that ergonomic standards will impact lost work days, direct costs, and disability.

Dealing with the politics of unions, purchasing, and those not understanding the on-site model was and has been continual. The model discussed throughout this book proposes solutions to get you started in preplacement/new hire examination, on-site job assessments, ergonomics, industrial rehabilitation, case management, cost savings, outcome reporting, and educating others about on-site services. If the goal is to return injured/ill workers to the jobs they held prior to the work-related incidents, then the politics seem like a small hurdle in comparison. The on-site concept is unique, but the three-legged stool of quality, cost, and service must balance. If one of the legs of the stool falls, they all fall. For example, one cannot have good quality without good service. I have been pleasantly surprised by the growth and positive benefits recognized by many chief executive officers, physicians, managers, and patients/workers.

The success of the on-site model has been achieved over 12 years. It is absolutely necessary to understand and focus on both the worker and the employer when practicing occupational health and industrial rehabilitation on site in industry. Employee/Patient education, patient rights and responsibilities, and just-in-time delivery of supplies to meet the mission and vision of the end product—returning the injured/ill worker to productive and gainful employment—will be a challenge for many years to come.

APPENDIX 1: RETURN TO WORK STATISTICS FOR BACK CASES AT AN ASSEMBLY PLANT

Patient name	S.S.N	Diagnosis	Date of Prescription	Date of Evaluation	Date of Discharge	No. of Treatments	Med-X	Work status
		Sciatica	8/19/97	9/2/97	9/24/97	4		Sick Leave
		Chronic Low Back Pain	1/9/98	1/13/98	1/20/98	4	X	Sick Leave
		Acute Lumbar Strain	3/27/97	4/3/97	4/25/97	11		Working
		Low Back Pain	1/29/98	1/29/98	2/10/98	2		Working
		Status Post Herrington Rod	1/16/97	1/16/97	1/16/97	1		Working
		Left SI Dysfunction	7/31/97	8/5/97	8/15/97	6		Working
		Low Back Pain	11/10/97	11/11/97	2/10/98	10	X	Working
		Herniated Lumbar Disc	4/1/97	4/29/97	10/15/97	12	X	Working
		Lumbar Myositis	1/21/98	1/21/98	2/18/98	2		Working
		Lumbar OA/DDD	1/22/98	1/29/98	2/11/98	4		Working
		Low Back Pain	1/31/97	1/31/97	2/5/97	3		Working
		S/P Lumbar Laminectomy	6/12/97	6/16/97	8/14/97	14	X	Working
		Lymbar Myositis	1/31/97	1/31/97	1/31/97	1		Sick Leave
		Low Back Pain	5/19/97	6/9/97	6/20/97	5		Working
		Low Back Pain	10/30/97	11/4/97	11/6/97	2	X	Working
		Chronic Low Back Pain	2/13/97	2/19/97	5/19/97	28		Sick Leave

		Back Pain	11/7/97	NA	11/17/97	0		Sick Leave
		Back Pain	12/17/97	12/18/97	12/22/97	3		Working
		Low Back Pain	9/22/97	10/2/97	10/3/97	1		Working
		Back Pain	2/12/98	2/16/98	2/20/98	3		Working
		Low Back Strain	4/8/97	4/14/97	5/28/97	8		Sick Leave
		Lumbar Spondylosis	8/6/97	8/12/97	9/5/97	9		Sick Leave
		Right Radiculopathy	11/13/7	11/20/97	1/22/98	10		Working
		Lumbar Myositis	1/26/98	1/28/98	2/9/98	6		Sick Leave
		Chronic Low Back Pain	9/30/97	10/1/97	10/17/97	6	X	Working
		Lumbar Pain	6/25/97	6/25/97	6/27/97	3		Working
		Lumbar Contusion	3/17/97	3/26/97	4/22/86	12		Working
		Lumbar Myositis	11/3/97	11/3/97	12/4/97	9		Working
		Acute Low Back Strain	3/25/97	4/15/97	5/16/97	12		Sick Leave
		Low Back Strain	3/6/97	3/12/97	5/9/97	20		Working
		Lumbar Myositis	9/8/97	9/8/97	NA	1		Working
		Chronic Low Back Pain	2/28/97	3/24/97	5/6/97	16		Working
		Sciatica	8/26/97	9/8/97	11/21/97	24		Sick Leave
		Low Back Pain	10/20/97	10/23/97	11/15/97	3		Sick Leave
		Low Back Pain	2/21/97	3/4/97	3/21/97	5		Working
		Low Back Pain	9/30/97	10/7/97	10/24/97	9		Sick Leave
		L/S Strain	9/29/97	9/30/97	10/23/97	0		Working
		Acute Back Pain	6/9/97	6/11/97	6/25/97	8		Working

APPENDIX 1—Continued

		Low Back Pain	9/16/97	9/18/97	9/30/97	5		Working
		Low Back Pain	1/30/97	1/30/97	2/10/97	6		Working
		Lumbar Strain	6/4/97	6/6/97	8/11/97	9	X	Working
		L/S Strain	1/30/97	3/13/97	6/20/97	12	X	Sick Leave
		Lumbar Strain	10/24/97	10/28/97	11/21/97	4	X	Sick Leave
		Low Back Pain	12/10/97	12/12/97	2/4/98	1		Working
		Lumbar Strain	5/12/97	5/19/97	8/1/97	19		Sick Leave
		Lumbar Myositis	6/24/97	7/16/97	8/8/97	9		Working
		L5/S1 DDD	10/20/97	10/21/97	11/6/97	9	X	Working
		Lumbar Strain	11/20/97	11/26/97	12/14/97	4		Sick Leave
		Low Back Pain	2/2/98	2/6/98	2/20/98	6		Working
		Acute Sciatica	10/14/97	10/23/97	10/24/97	1		Sick Leave
		Chronic DDD	2/13/97	3/4/97	4/3/97	10		Sick Leave
		Low Back Pain	5/16/97	5/20/97	9/8/97	22	X	Working
		Low Back Strain	9/18/97	9/18/97	10/2/97	7		Working
		Lumbar Spondylosis	2/17/97	3/6/97	3/12/97	2		Working
		Chronic Low Back Pain	1/13/98	1/15/98	2/24/98	6		Sick Leave
		Chronic Low Back Pain	2/26/97	3/21/97	4/14/97	11		Sick Leave
		Acute Lumbar Myositis	9/15/97	9/15/97	12/5/97	11	X	Working
		Spondylolisthesis	2/17/97	2/26/97	4/11/97	13		Working
		Back Strain	12/12/97	12/16/97	2/10/98	4		Working
		Low Back Strain	12/3/97	12/4/97	12/17/97	6		Working

		Lumbar Laminectomy	9/30/97	10/1/97	10/20/97	7	X	Working
		Lumbar Laminectomy	1/22/98	1/26/98	2/23/98	10		Working
		S/S SI Joint	12/17/97	12/19/97	1/21/98	6	X	Sick Leave
		Lumbar Myositis	7/14/97	8/27/97	8/29/97	3		Sick Leave
		Low Back Pain/Sciatica	9/23/97	10/7/97	12/15/97	12		Working
		Acute Lumbar Strain	8/12/97	8/12/97	8/19/97	4		Working
		Chronic Low Back Pain	2/3/97	2/10/97	2/20/97	4		Working
		Lumbar Strain	11/21/97	11/24/97	12/23/97	13	X	Sick Leave
		Lumbar Discomfort	8/4/97	8/12/97	9/19/97	12		Sick Leave
		Low Back Pain	2/20/97	2/20/97	3/25/97	6		Working
		Back Pain	11/20/97	11/21/97	2/19/98	15		Working
		Low Back Strain	12/5/97	12/8/97	1/21/98	10		Working
		Lumbar Strain	10/2/97	10/3/97	10/31/97	11		Working
		Lumbar Myositis	9/9/97	9/11/97	9/15/97	3		Working
		L/S Strain	7/17/97	7/17/97	7/24/97	5		Working
		Back Pain	12/10/97	12/10/97	2/12/98	10		Working
		Acute Lumbar Myositis	8/1/97	8/11/97	8/19/97	5	X	Working
		Low Back Pain	6/2/97	6/19/97	7/14/97	4	X	Working
		Acute Lumbar Myositis	9/9/97	9/16/97	9/25/97	4		Working
		Low Back Strain	2/17/97	2/20/97	2/20/97	1		Sick Leave
		Right L5 Radiculopathy	11/21/97	12/15/97	12/23/97	2	X	Working
		S/P Lumbar Laminectomy	9/30/97	10/2/97	10/8/97	2		Working

APPENDIX 1—Continued

		Low Back Pain	2/11/97	2/13/97	3/26/97	6		Working
		Sacroilitis	1/12/98	1/13/98	1/30/98	8		Working
		Chronic Low Back Pain	1/28/98	2/2/98	2/27/98	12		Working
		Acute L/S Strain	3/25/97	3/25/97	4/28/97	16		Working
		Back Pain	11/17/97	11/18/97	11/18/97	1		Working
		Lumbar Strain	10/20/97	10/22/97	12/15/97	5	X	Working
		Back Strain	10/30/97	10/30/97	12/12/97	14	X	Working
		Lumbago/Sacroilitis	1/28/97	2/12/97	2/28/97	6		Working
		Chronic Low Back Pain	5/27/97	5/28/97	6/10/97	5		Working
		Low Back Strain	4/8/97	5/13/97	6/20/97	14		Working
		Low Back Pain	3/25/97	4/22/97	7/18/97	24	X	Working
		Low Back Pain	2/19/98	2/20/98	3/4/98	6		Working
		Back Pain	3/13/97	3/25/97	5/30/97	25	X	Working
		Acute Lumbar Spasm	1/31/97	2/5/97	2/11/97	6		Working
		Low Back Pain	1/26/98	NA	2/5/98	0		Working
		Acute Lumbar Strain	8/15/97	8/20/97	9/5/97	1		Working
		Acute Low Back Strain	2/24/97	3/19/97	3/26/97	5		Working
		Lumbar Strain	1/9/98	1/9/98	1/16/98	3		Working
		Back Pain	11/5/97	11/11/97	1/22/98	11	X	Sick Leave
		Low Back Pain	5/1/97	5/2/97	6/24/97	5		Working
		Right L/S Myositis/ Sciatica	4/15/97	4/16/97	5/5/97	6		Working

		Lumbar Strain	2/14/97	2/20/97	4/14/97	14	X	Working
		S/P Lumbar Laminectomy	4/8/97	5/20/97	6/24/97	3		Working
		Radiculopathy/ Low Back	11/5/97	11/5/97	1/28/98	11	X	Working
		Lumbar Strain Sciatica	4/17/97	4/17/97	9/3/97	30	X	Working
		Acute L/S Torsion	8/26/97	8/26/97	9/25/97	5		Working
		Back Strain	11/17/97	11/17/97	12/17/97	13	X	Sick Leave
		Low Back Pain Radiculopathy	8/4/97	8/4/97	9/11/97	5		Working
		Back Pain	2/21/97	2/28/97	5/22/97	24	X	Working
		Right SI Pain	3/11/97	3/14/97	4/22/97	5		Working
		Low Back Pain	6/20/97	6/20/97	6/27/97	4		Working
		T11-L2 Spasm	8/29/97	8/29/97	9/11/97	5		Working
		Lumbar Strain	5/23/97	5/27/97	6/13/97	9	X	Working
		Low Back Pain/ DDD	11/13/97	11/18/97	11/26/97	4		Sick Leave
		Acute Lumbar Spasm	8/14/97	8/14/97	8/28/97	10		Working
		Low Back Pain	9/30/97	10/8/97	10/20/97	6		Working
		L/S Strain	7/21/97	7/21/97	7/21/97	1		Working
		S/P Lumbar Laminectomy	7/29/97	8/1/97	8/22/97	6		Sick Leave
		Chronic Low Back Pain	9/15/97	9/15/97	12/23/97	11	X	Working

APPENDIX 1—Continued

		S/P Lumbar Laminectomy	7/23/97	7/23/97	8/1/97	4	X	Sick Leave
		L/S Radiculopathy L/S Spondylosis	10/7/97	10/10/97	2/10/98	25		Working
		Dorsal Myositis	9/26/97	9/29/97	10/3/97	2		Working

Number of Discharges = 124; MedX Discharges = 31.
Number of Patients = 118 (six patients received two courses of treatment and were discharged twice).
Based on 124 discharges for back cases:

29	23.39%	remained on sick leave after discharge
95	76.61%	ere physically able to return to work or continue working
124	100.00%	

MedX results:
25% of all lumbar cases were placed on the MedX.
8 of 31 or 25.8% remained on sick leave after discharge.
23 of 31 or 74.2% were physically able to return to work or continue working.
Physical therapy results:
21 of 93 or 22.6% remained on sick leave after discharge.
72 of 93 or 77.4% were physically able to return to work or continue working.

APPENDIX 2: ON-SITE INDUSTRIAL REHABILITATION: 12-MONTH STUDY OF PERCENT BODY FAT LOSS THROUGH USE OF THE FITNESS CENTER

This random study was performed over a 12-month period surveying 35 out of an average of 100 exercise participants percentage body fat. Percent body fat is a measurement taken to determine what percentage of body weight is composed of fat in relation to that composed of fat-free tissue. The advantage of decreasing body fat over a prolonged exercise program are (1) a decreased incidence of cardiovascular disease, (2) an increase in life expectancy, (3) decreased problems with muscles and joints, (4) a decreased incidence of diabetes, (5) lower blood cholesterol levels, and (6) a decreased incidence of other health-related problems.

Gain or loss of body fat	Number of persons[a]	Percentage
Gain of 0–2%	3	8
Loss of 0–1%	7	21
Loss of 1–2%	6	17
Loss of 2–3%	8	23
Loss of 3–4%	6	17
Loss of 4–5%	2	6
Loss of 5–6%	3	8

[a] The total number of participants for this study was 35.

APPENDIX 3: NET REVENUE GROWTH VS. 1989: DIRECT COMPANY CONTRACTS

	1990	1991	1992	1993	1994	1995	1996	1997	1998
Growth vs. 1989, %	56.3	163.6	268.5	441.7	414.3	491.5	382.0	902.7	2587.7
Annual growth, %	56.3	68.6	39.8	47.0	−6.1	15.0	−18.5	108.0	168.0

APPENDIX 4: DEVELOPED SYSTEM PROCEDURES TO SUPPORT THE ISO-9000 QUALITY MANUAL

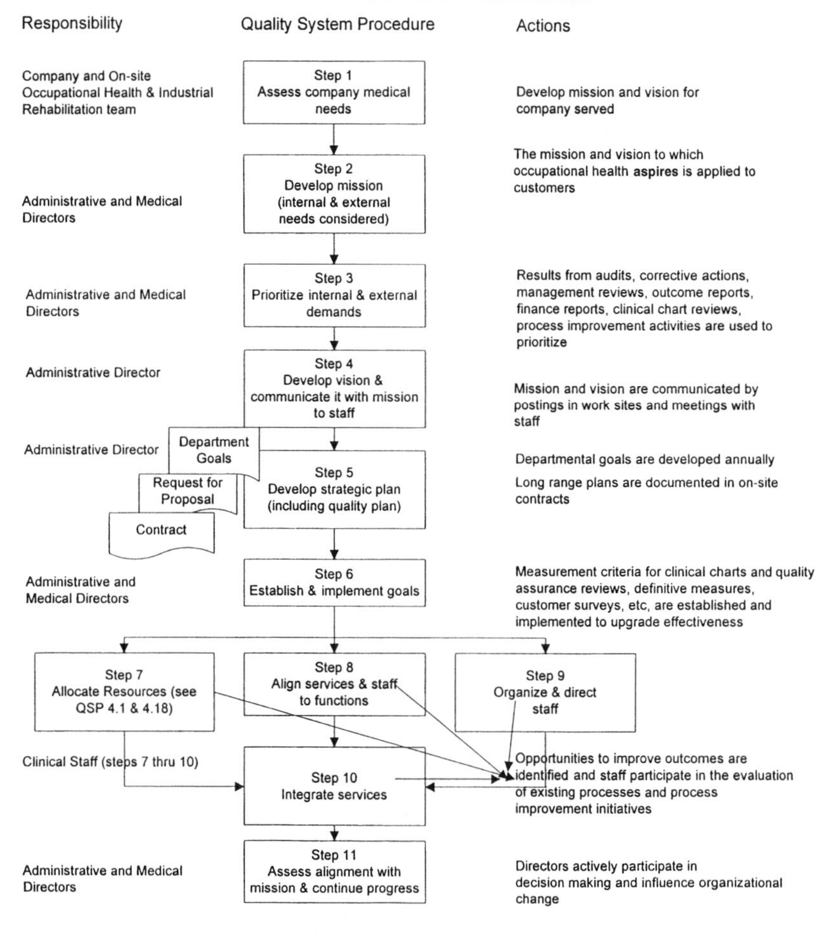

References

1. American Medical Association. Public Opinion on Health Care Issues. Chicago: American Medical Association, August 1997a.
2. American Medical Association. Physician Opinion on Health Care Issues. Chicago: American Medical Association, August 1997b.
3. American Physical Therapy Association, McCann, KB. Letter. March 13, 1995.
4. Battie MC, Bigos SJ, Fischer LD, et al. Isometric lifting as a predictor of industrial back pain reports. Spine, 1989; 14:851–856.
5. Bodenheimer MD. Health Policy Report. The American health care system. The movement for improved quality in health care. N Engl J Med 1999; 340:488–492.
6. Cattledge GH, Schneiderman A, Stanevich R, Hendricks S, Greenwood J. Nonfatal occupational fall injuries in the West Virginia construction industry. Accident Anal Prev 1996; 28(5):655–663.
7. Chassin MR, Galvin RW. The urgent need to improve health care quality. JAMA 1998; 280:1000–1005.
8. Grumbach K, Anderson GM, Luft HS, Roos II, Brook R. Regionalization of cardiac surgery in the United States and Canada: Geographic access, choice, and outcome. JAMA 1995; 274:1282–1288.
9. Hannan EL, Kilburn H, Jr, O'Donnel JR, et al. A longitudinal analysis of the relationship between in-hospital mortality in New York State and the volume of abdominal aortic aneurysm surgeries performed. Health Serv Res 1992; 27:517–542.
10. Herrin G. Why Ergonomics? Occupational Ergonomics Workshop. The University of Michigan Center for Occupational Health & Safety Engineering, April 23 & 24th, 1992.
11. Jollis JG, Peterson ED, DeLong ER, et al. The relation between the volume of coro-

nary angioplasty procedures at hospitals treating medicare beneficiaries and short-term mortality. N Engl J Med 1994; 331:1625–1629.

12. Karp HR, Flanders WD, Shipp CC, Taylor B, Martin D. Carotid endarterectomy among Medicare beneficiaries: A statewide evaluation of appropriateness and outcome. Stroke 1998; 29:46–52.
13. Leigh JP, Markowitz SB, Fahs M, Shinn C, Canrigan PJ. Occupational injury and illness in the United States. Estimates of costs, morbidity, and mortality. Ach Intern Med 1997; 157:1557–1568.
14. Mahone D, Burkhart J. Ergonomics: Prevent back injuries or pay later. National Underwriter Property and Casualty, May 1992, p. 11.
15. Mayo Clinic. Back care: What's behind back pain and what you can do to prevent it. Mayo Clin Health Lett 1994; Feb(Suppl):1–8.
16. Mooney V, Kron M, Rummerfield P, Holmes B. The effect of workplace based strengthening on low back injury rates: A case study in the strip mining industry. J Occup Rehabil 1995; 5(3):157–167.
17. Moore JD, Jr. HEDIS 2000 Emphasizes Prevention. Modern Healthcare 1999; 29(8): 32.
18. Nelson BW, O'Reilly E, Miller M, Hogan M, Wegner JA, Kelly C. Focus on the spine. The clinical effects of intensive, specific exercise on chronic low back pain: A controlled study of 895 consecutive patients with 1-year follow up. Orthopedics 1995; 18(10):971–981.
19. Williams JM, Higgins D, et al. Work-related injuries in a rural emergency department population. Acad Emerg Med 1997; (April):277–281.

Suggested Readings

Adler AL. If at first you don't succeed . . . GM takes another whack at product development structure. Ward's Auto World 1995; (Sept.):65–67.

Brody J. Carpal tunnel syndrome: Some new treatments for it. The Ann Arbor News 1995; (March 3):F-3.

Carlquist SA, DeHart JP. Developing and marketing a successful work program. The American Occupational Therapy Association, Inc., Work Programs, Special Interest Section Newsletter 1988; 11(4):1–3.

Carson R. Machine and tool safety. Ergonomic advantages of air-powered cutting tools. Occupational Hazards 1994; (June):139–140.

Chapel TJ, Strange PV. Is altruism killing prevention? When health systems taken on the role of public health provider. Healthcare Forum 1997; (Sept.–Oct.):46–50.

Chassin MR. Quality of health care. N Engl J Med 1996; 335(14):891–894.

Danna V. On the road to recovery and better health. Occupational health leaves the hospital and enters the workplace. Healthcare Benefits Rev 1999; 4(Jan.):29, 31, 33.

Dyre JH. Chrysler transplanted Japanese-style supplier relations to the competitive soil of the United States. How Chrysler created an American Keiretsu. Harvard Business Rev 1996; (July–Aug.):42–56.

Galusha J. Applied ergonomics: Systematic, criteria-based intervention. Occupational Hazards 1994; (Jan.):67–70.

Galvin RS. What do employees mean by "value"? Integrated Healthcare Report 1998; (Sept.–Oct.):1–15.

Jenson MD, Brant-Zawadzki MN, Obuchowski N, et al. Magnetic resonance imaging of the lumbar spine in people without back pain. N Engl J Med 1994; 331(2):69–73.

McGrail MP, Jr, Tsai SP, Bernacki EJ. A comprehensive initiative to manage the incidence and cost of occupational injury and illness. J Occupational Environ Med 1995; 37(11):1263–1268.

Meighan S. Where have all the primary physicians gone? A Socratic discourse. Health Care Manage Rev 1995; (Summer):64–67.

Mont D, Burton JF, Jr, Reno V. Workers' compensation: Benefits, Coverage, and Costs, 1996 New Estimates. National Academy of Social Insurance, Washington, D.C., March 1999.

Nickel C, Yangouyian S. Disabled doesn't mean deficient. Low-cost workplace modifications can help temporarily or permanently disabled employees. Occupational Health Safety 1996; (April):52, 53, 69.

Nockels R, Mandel S, Mitchel L, Morlock R, Shaffery C, Rauzzino M, Rosenblum M. Application of disease management principles to spinal disorders: Early experience in group practice managed care environment. AANS/CNS (Spinal Peripheral Nerves Section). Abstract. 1999.

Ott K. At-work health programs cut lost days for two companies. Crains Detroit Business 1998; 14(8).

Steckol KF. Fast forward. ASHA leaders imagine where we'll be in the year 2005. ASHA J 1995; (Jan.):50–56.

Treatment of back pain: Outpatient service charges, 1993 (recap 1993 outpatient charges for back pain treatment). Statist Bull 1995; (July–Sept.).

Wiley G. Outcome and the future of rehab. The need to reduce costs while producing beneficial results is a driving factor behind the trend toward measuring outcomes in rehabilitation. Rehab Manage 1992; (June–July):123–125.

Winslow R. The payers, firm hands, corporate coalitions have been a major force in keeping costs down. What's their current game plan? Wall Street 1997; (Oct. 23):R19.1.

Wojcik J. Work comp claims deserve special treatment, goals differ from group health arena. Business Insurance 1994; (April 25).

Index